Communication and Global Society

PETER LANG
New York • Washington, D.C./Baltimore • Boston • Bern
Frankfurt am Main • Berlin • Brussels • Vienna • Oxford

Communication and Global Society

Edited by
Guo-Ming Chen
and William J. Starosta

PETER LANG
New York • Washington, D.C./Baltimore • Boston • Bern
Frankfurt am Main • Berlin • Brussels • Vienna • Oxford

Library of Congress Cataloging-in-Publication Data

Names: Chen, Guo-Ming, editor. | Starosta, William J., editor.
Title: Communication and global society /
edited by Guo-Ming Chen and William J. Starosta.
Description: New York: Peter Lang, 2000.
Includes bibliographical references and index.
Identifiers: LCCN 99048693 | ISBN 978-0-8204-4865-7 (paperback: alk. paper)
Subjects: LCSH: Communication, International. | Pluralism (Social sciences).
Internet (Computer network). | Intercultural communication.
Interpersonal communication.
Classification: LCC P96.I5.C657 2000 | DDC 302.2—dc21
LC record available at https://lccn.loc.gov/99048693

Bibliographic information published by **Die Deutsche Nationalbibliothek**.
Die Deutsche Nationalbibliothek lists this publication in the "Deutsche Nationalbibliografie"; detailed bibliographic data are available on the Internet at http://dnb.d-nb.de/.

Cover design by Lisa Dillon

The paper in this book meets the guidelines for permanence and durability of the Committee on Production Guidelines for Book Longevity of the Council of Library Resources.

29 Broadway, 18th floor, New York, NY 10006
www.peterlang.com

Printed in the United States of America

TABLE OF CONTENTS

PREFACE

When new means of communication are brought to bear on existing social institutions, change may be expected in both. Social institutions are reshaped by circumventing the nation-state, by provoking dialogues among people who are forbidden by their governments to talk with one another, and by building up virtual communities that transcend existing boundaries. Alternatively, communication technologies respond to the context of which they are a part.

Persons within particular contexts trust new information technologies, or not; they feel that their conversations are private or that they are not; they place great hope on interaction across cultures or not; they open themselves as listeners to new ideas or not.

We have assembled some of the most interesting writers who have searched for relationships among new forms of communication and existing practices and institutions. Some of the writers stress that changes in communication will provide a severe blow to the idea of the nation-state, while others stress the power to create virtual communities that will harmonize themselves within existing institutional formats. Whereas some writers see the consequences of global communication as provoking conflict, others see it as harmonizing communities and nations. Though the informational future has already arrived in some measure, it remains notoriously difficult to write contemporary history.

Many writers of this collection have interacted on list-serves and at national communication conferences, and have maintained an ongoing dialogue over a period of some twelve years. Others were invited to contribute on the basis of their known research orientations. The resultant mix of ideas, for its differences, returns to some common themes. These themes permit us to arrange the chapters into broad groupings.

We express our excitement about the ideas covered in this collection. The breakdown between disciplinary perspectives, the dissolving of walls between mass and interpersonal communication, the circumventing of nation state, the reinforcements of nationalisms, the transcending of state, the maintenance of virtual communities of persons widely separated by geography, all offer new grist for the mills of scholars and futurists.

We would like to thank the authors who contributed to this volume and worked many months to write and revise their chapters. We are indebted to the editorial staff at Peter Lang Publishing for their consistent support and suggestions on the completion of this volume. We are also grateful to the Department of Communication Studies at the University of Rhode Island and Howard University for providing support on various aspects of this project.

Guo-Ming Chen
William J. Starosta

CHAPTER ONE
Communication and Global Society: An Introduction

Guo-Ming Chen
University of Rhode Island
William J. Starosta
Howard University

A GLOBALIZING WORLD

We live in a globalizing society. A web of communication and transportation technology has shrunk the world so that the global interdependence for people and cultures has become a norm of life. Internationally, people from different cultures increasingly and inevitably encounter each other during travel, study, and business transactions. Domestically, especially in the United States, the movement toward multiculturalism or cultural diversity has become a cascade brought by the rapid pace of immigration and the expectation by national minorities for equitable treatment. In other words, the globalization trend influences every aspect of our life individually, interpersonally, and in organizations.

As the world becomes more interdependent and interconnected, the nation-state becomes more culturally heterogeneous. In consequence, increasingly frequent face-to-face interaction among people from different cultural, ethnic, social, and religious backgrounds demands that we develop a global mindset by which we try to see things through the eyes of others and add their knowledge to our personal repertoires. Such a global mindset can only be maximized through increased intercultural communication competence. Only through global cultural competency can people from different backgrounds communicate effectively and successfully in the upcoming global society (Chen & Starosta, 1996, 1998). Skill in global competency fosters within individuals multiple, simultaneous identities in terms of culture, ethnicity, race, religion, nationality, and gender. It transforms us from monocultural persons into multicultural persons and functions to nourish a human personality in which we are aware of our multiple identities and are able to maintain a multicultural coexistence in order to develop a global civic culture (Adler, 1982). Moreover, global competency enables us to demonstrate tolerance and mutual respect for cultural differences as a mark of enlightened national and global citizenship on the individual, social, business, and political institutional levels (Belay, 1993; Morgan, 1988). Therefore, global competency becomes a critical ability for adjusting to the demands of the new millennium.

In order to foster the ability of global competency, it is important to first understand that a global society demands of us new modes of organization and thinking (Rhinesmith, 1993). If ever it was possible to remain self-absorbed in local, rural culture, that day is passing soon, or it has already passed us by. Bit by bit, we became enmeshed in matters external to our village. We visited other countries as tourists, we went abroad to seek out forgotten relatives, we studied abroad, we sought larger markets for our products and services, we participated overseas in military or in nongovernmental organizations, or we took intensive coursework in ethnic music or in the cultural expression of other civilizations present and past. With each movement toward broader cultural understanding, we left bits of our village culture behind. With each new encounter, we began to prefer alternative styles of food, architecture, music, and literature. Each business and professional contact engrained in us a desire to culturally retool so that we could better function with those of unfamiliar agencies and businesses. Our awareness of other possibilities, like communication, proved to be a cumulative process. Once encountered, once learned, once transacted, our increased awareness became irreversible.

Our increased awareness gradually brings to an end cultural separation and naïveté. We started to compare standard English with Ebonics; Japanese theories of business with American; North American pragmatism with Indian idealism; monotheism with polytheism; patriarchy with matriarchy; saying "no" to mean "no" versus saying "probably yes" to mean "no." Time frames became unclear. Messages abounded with unexpected nuances. Benefits expected from the development of new markets had to be weighed against the vagaries of new bureaucracies, new relationships between governments and businesses, attitudes, ethnocentrisms, animosities. Sometimes the answer to the influx of new cultural knowledge was to retreat into isolation and prejudice, to expel foreign expatriates, to "cry foul." Gains in cultural awareness came at the price of new frictions and resistance (Richmond, 1994).

But cultural awareness could not be undone. The first successful multinational corporations made a compelling case against business isolation.

The brain drain that impoverished some nations of their finest minds enriched other nations with Nobel laureates and with honored technicians and doctors. Before many adults had learned to turn on a computer their children had struck up multicultural, multiethnic friendships on the Internet. Ethnic animosities came to be weighed against the possible benefits of participating in the expanding world of international trade. Citizen-to-citizen contact prospered even in the face of governmental restrictions and sanctions.

Our contact with others unsettled our personal sense of geography and community. We had to ponder which foreign language training, if any, we might need to some day gain productive employment. We had to decide which academic major would provide a sense of achievement and fulfillment in this changing world. We asked which foreign markets were natural for the creation of free trade zones. We began to identify similarities in conditions among specific countries that suggested the need to coordinate governmental policy. Our primary identity within our village became a racial memory, a place to which to return for weekend visits between forays into places outside the village. Some of our forays were physical, using improved modes of transportation. Most, however, were virtual, mediated by some mode of synchronous or asynchronous communication (Cooper & Selfe, 1990; December, 1996; Harasim, 1993; Ma, 1994).

With our sense of physical and psychic displacement came the need to reconsider and to redefine community. Could any mode of virtual connection between physically-distant individuals provide a complete sense of proximate, intimate co-involvement? Could interactions across distance on a restricted number of topics possibly replace the diversity that earlier characterized our villages? Or should it?

In the wake of gains at the level of virtual community, we realized threats to our cultural uniqueness. We were forced to ask: Would gains in globalization signify the end to village identity, to our definition as geographically-situated, linguistically-unique, culture-rich individuals (Naisbitt, 1994)? Further, if we sacrificed our village moorings for something more global, what would we talk about? To which virtual organization or entity would we owe allegiance (Adler, 1997)? Would we need new values, new powers and skills of tolerance and adaptation and organization? Increasingly, unremittingly, almost intangibly, our questions became those of communication and global society. Thus, we ask: How can communication scholars make contributions through their research to help people understand these questions and better adapt to the globalizing society?

A RESEARCH AGENDA

The above discussion offers some research concerns on communication and global society: (1) continuity and change of identity in global community, (2) the emergence and impact of global media, and (3) the study of the nature and forms of interaction in global society.

Continuity and Change of Identity in Global Community

The emergence of global society does not necessarily mean that a sense of community will follow. A new sense of community must be cultivated in order to build a cohesive global society. In this new global community no one can exist without the other. The community is inclusive. It manifests "allness" that draws together those who differentially experience race, ethnicity, gender, spirituality, creed, and emotional release. Community requires commitment. It demands that citizens participate equally and fairly in the construction of global society. This community seeks voluntary consensus. It pursues agreement among all the members by acknowledging and discussing differences with a view toward acknowledgment and celebration instead of ignoring or denying of the other's standpoint (Peck, 1992). In other words, this global community is an "epistemic community" defined not by members' racial, political, or geographic boundaries but by common beliefs, values, and symbolic schemas (Thayer, 1987).

This new sense of "epistemic community" compels us to reassess the concept of global citizenship. Should global citizenship be equated with nationality and membership of the local community, or will citizenship be reckoned on the basis of virtual, psychoemotional identifications? In other words, how do we negotiate our status in three levels of group affiliation: community membership, national citizenship, and global citizenship? Obviously, affiliation with a particular national or a local community should not deny equal status to those of other societies and communities through ignorance or culturally unequal power relationships and exploitation (Boulding, 1988; Lynch, 1992). Can human beings promote voluntary pluralism by integrating different cultural identities and interests to build a global civic culture? If so, how?

We assert that through communication, citizens of the global society begin to develop "communication realities" (Thayer, 1987) in which ideas, beliefs, preferences, qualities, evils, and ideals can be talked about, and through which members can use symbols to recreate themselves and to define why they are a particular kind of human in the global society. Safe space must be found to discuss issues of ethnicity, race, power, gender, and psychological orientation. Ultimately, global citizens establish their own conscious identities and a total social environment in which they coexist and rationalize their own and others' actions to reach the global epistemic community.

Global society then represents a space in which many cultures are integrated and continuously (re)defined. With each new transaction, individuals are made to choose among numerous possible identities. How individuals negotiate and co-create cultural identity through communication in the culturally diverse global society becomes a critical issue (Collier & Thomas, 1988). In other words, in the global society individuals must learn to nourish a human personality in which they are aware of their multiple identities and are able to maintain a multicultural coexistence (Belay, 1993). Naisbitt (1994) used "global paradox" to describe the potential problem we face in the globalizing process. Globalization not only demands an integration of cultural diversity in the global community, but at the same time also reflects people's needs to develop a strong self or cultural identity(ies). The dialectic tension between cultural identity and cultural diversity poses a great dilemma for future society. Global society will be a mirage if individuals fail to establish an equilibrium in the dialectical interaction between cultural identity and cultural diversity. Thus, the first and foremost issue for scholars to investigate the globalizing society is to examine the emergence of a new sense of community that reflects the dialectical relationship between cultural identity and cultural diversity.

The Emergence and Impact of Global Media

The virtual environment of media not only functions to connect people with one another and to link people to the larger global society, it also plays a determinative role in the political process by creating a "public sphere" in which issues of importance to a global community are discussed and debated (Herman & McChesney, 1997). Media lead to citizen participation in global community life. They present a new vista by potentially disseminating information to every corner (subject to availability of new media) that in turn facilitates dialogue among diverse cultures in the global society (Stevenson, 1994). Moreover, the global media may further reshape the pattern of intercultural communication by creating common meanings and a reliance on persons we may have never met in our life (Porter & Samovar, 1994). For instance, more and more people begin "to talk and type their way daily into a web of mediated intercultural interaction" (Chen & Starosta, 1998, p. 4). According to Chen (1998) and Hart (1998), electronic mail via Internet has become the most common form of computer-mediated communication today. The electronic mail system allows people to communicate all over the world in a few seconds in one-to-one, one-to-many, or many-to-many formats. It provides us a powerful and non-conventional way for dialogue and collaboration that counters the constraints of time and space (Garnsey & Garton, 1992).

Unfortunately, media also lend themselves to commercialization, commoditization, competition, and other ideological traps in the process of globalization. The ideology of the commercial model of media that is controlled by advertisers can subvert the democratic order and gradually erode the public sphere. In addition, Kim (1998) points out that due to the global unequal flow of media, cultural imperialism may also emerge in new interactive electronic media. How media function to preserve indigenous cultural-political space and to cope with local and national resistance will decide the future of global society. Thus, the examination of the nature and function of global media, especially the Internet, whether or not such interaction is essentially interpersonal, and the potential impact of media on human consciousness in the global society is vital for study by communication researchers.

The Study of Nature and Forms of Interaction in Global Society

The globalization trend not only impacts every aspect of our personal and social life, but also leads to a higher level of interdependence and interconnectedness between individuals, organizations, and governments in different cultures. Thus, in order to live and cooperate harmoniously and productively with one another in the global society, it is indispensable for us to increase our knowledge and skills in intercultural interactions. In other words, we need to acquire communication competence in the global context that will produce in us understanding, tolerance, and mutual respect for cultural differences.

More specifically, as enlightened global citizens, we must envisage changing world trends and engage in the process of regulating the change within a context in which diversity and cultural differences are valued and balanced. We need not only cultivate the ability to map our own and another's cultures, but also to monitor ourselves so we may harmonize our identity with that of our continuity by extending personal attributes, such as sensitivity and open-mindedness, to build a bridge across personal, social, and cultural gaps. In addition, to adjust ourselves to the changes and new patterns of interaction in the global society, we need to learn those behavioral requirements, such as flexibility, active listening, and interaction management, that enable us to negotiate the multiple meanings and manage complexity and conflicts in the global context.

Thus, communication plays a key role in the process of developing a global civic society. Only through communication can the foundation for global society be built. It is then obligatory for scholars to investigate human interactions in the process of forming a new sense of community that reflects the dialectical relationship between identity and diversity in the global context. The exercise of the faculties we describe may lead in unexpected directions wherein nation becomes obsolete and identities shift with each new context. The changes we anticipate will occur in a context where some persons with substantial power may not recognize the need to build a sense of global community. Interactions leading to globalization will steer a course between dialogue and debate, nationality and ethnicity, self and community, geographic proximity and virtual community, legal constitution and experienced subjectivity. Our investigations of global community will reflexively shape the world that is unveiled.

AN OVERVIEW OF THE BOOK

The three categories of research agenda for communication and global society serve as the three parts of this book. The 16 chapters included in this book represent a diverse theoretical and methodological orientation for the study of communication and global society. They add to the growing literature in this line of research by investigating the problem regarding cultural diversity vs. cultural identity; globalization vs. localization; sense of global community; the impact of global media such as TV and Internet, global ethical communication, and conflict, listening, human understanding, and relationship in the global context.

Part I deals with continuity and change of identity in global community. In addition to the introduction chapter, four chapters are included in this part. Rueyling Chuang's "Dialectics of Globalization and Localization" synthesizes and brings to light the different views on the ramifications of globalization in the post/modern era. The chapter delineates theorists' various concerns with regard to globalization such as globalization as essentially Westernization, Europeanization or Americanization that might lead to cultural imperialism It also parallels other theorists' (e.g., Featherstone, 1990, 1991; Robertson, 1992, 1995) contention that globalization is a representation of postmodernity in which marginal voices are privileged, and diversity is valued. The chapter further explicates the dialectic contradictions between localization and globalization and the implications of a unified world culture as the consequence of globalization, as well as the homogenization and heterogenization of local cultures. Thus, the foci of the chapter include the dialectical process of cultural integration, cultural disintegration, and the effects of an emerging transnational culture, global culture, and a third culture.

Mei Zhong's "Dialectics of Identity and Diversity in the Global Society" explores the issue of a dialectic relation between global diversity and people's struggle with identity. Literature on the trend toward diversity in most societies, especially in the United States and China, is reviewed. The chapter also reviews identity literature in terms of formation at individual and social group levels. A dialectic relation between the two entities is demonstrated with application in theory, research, applications, and examples. Specifically, the tension between the two entities is demonstrated (1) in the theory of intercultural adaptation, both in theory and in practice, (2) in intercultural communication research in the area of cross-cultural communication competence, and (3) in real life situations such as in the controversial issue of national language(s) and preservation of local identities.

Hui-Ching Chang's "Reconfiguring the Global Society: 'Greater China' as Emerging Community" argues that citizens of global society not only must assure their identities but also extend themselves into multiple domains of existence. The author points out that this greater interdependence simultaneously heightens both cultural self-awareness and the urge for alternative alliances that transcend traditional national boundaries. In the wake of the Cold War, there has been a vacuum challenging the "citizens" of the global society to re-structure a new world order. This chapter therefore critically analyzes one such alternative formulation, brought about by an aggregation of Chinese articulated in talk of Chinese unification and the concept of "Greater China." Since the takeover of Hong Kong in 1997 by the People's Republic of China, the dynamics of the interlinkages between the PRC and Taiwan (ROC) have shifted. This chapter addresses questions concerning how Chinese unity may be achieved and how such unity may serve to redefine the Chinese sense of identity, to change the equilibrium of international politics, and to redefine the parameters of global society.

Finally, Jensen Chung's "The Challenge of Diversity in Global Organizations" argues that most problems and solutions in the literature concerning diversity are examined from the perspectives of the majority, rather than from those of the co-cultures. Because diversity is an inevitable issue and challenge in the global society, this chapter illustrates this irony by identifying the roots of assertiveness problems of a co-cultural group. Through interviewing Taiwanese American engineers in global corporations and holding focus group meetings with them, ten major themes from the transcripts are identified. The themes, organized along communication theories and personal conceptualizations, are compared with the assertiveness literature to show the need for modern organizations to pay attention to localization in the process of globalization. In other words, it shows that from the perspective of organizational culture, global organizations need to acknowledge diversity in their efforts of unifying corporate culture.

Part II deals with the emergence and impact of global media. Five chapters are included in this part. Ringo Ma's "Internet as a Town Square in Global Society" argues that the Internet has created a global town square rather than merely an arena for elite media programs. On this town square, the line between interpersonal communication and mass communication is blurred. Participants enjoy the freedom from being brainwashed by state-sanctioned media programs. The new digital medium makes gate-keeping more difficult while underrepresented groups are heard more frequently, so it is counter-hegemonic. The self-image projected and the reality constructed on the Internet demonstrate a new extension of our physical being and environment. The three characteristics of communication associated with the Internet, including demystification of exaggerated cultural differences, empowerment of the underrepresented, and creation of hyperself and hyperreality, imply the emergence of an internationally homogeneous and yet locally diverse society. They also demand the establishment of a new world order as a way to prevent negative impacts from the Internet.

Richard Holt's "'Village Work': An Activity-Theoretical Perspective Toward Global Community on the Internet" uses Yrjo Engestrom's activity theory to propose an alternative way of looking at global community on the Internet. It is demonstrated that current ideas about Internet community fail sufficiently to prioritize activity. Activity is central to Engestrom's theory as that which binds humans together in what Marx referred to as the paradox of production and consumption. It is argued that the activity of computer-mediated communication (CMC) is a response to a need state (as is all human activity), inextricably tying together six elements: who performs the activity (subject); the means by which it is accomplished (instrument); goal of the activity (object); the aggregate of individuals who share meaning with respect to the activity (community); guides for the activity espoused by the community (rules); and task specialization regarding the activity by individuals and groups who comprise the community (division of labor). A year-long corpus of postings to a public bulletin board, part of an ongoing worldwide teaching/learning project which involves high school students in discussion of utopian communities, is examined to reveal the various systems and subsystems of activity involved. Findings support the idea that studies of community, both computer-mediated and otherwise, will benefit from less emphasis on abstraction and more emphasis on sociohistorical specificity. It is concluded that the notion of global society should not be understood as simply a utopia in which participants can freely use the Internet to communicate, but in addition as activity engaged in by Internet users which necessarily and inevitably places constraints on their range of choices.

Guo-Ming Chen's "Global Communication via Internet: An Educational Application" reports the results of a 3-year experiment on global communication through e-mail debates among students in Denmark, France, Germany, Hong Kong, Turkey, and the United States. The author argues that in order to survive in this ever-shrinking global world we must learn to see through the eyes and minds of people from different cultures, and further develop a global mindset to live meaningfully and productively in the 21st century. To successfully use the new technology systems like e-mail for improving students' abilities in intercultural communication will prove to be a promising step toward achieving this goal. In this chapter a series of assessments and effects of Internet on global communication are reported. Problems encountered and suggestions for future experiments in global Internet exchanges are also discussed.

Gary Larson's "Globalization, Computer-Mediated Interaction, and Symbolic Convergence" points out that the Internet provides researchers an arena in which to view the enactment of—and resistance to—a global dialogue. In this chapter the author: (1) discusses a facet of Internet communication that retains a dynamic and at times even anarchistic character: the temporally synchronous discussion channels known as Internet Relay Chat, (2) describes the instrumentality and community of IRC, and (3) demonstrates how the theoretical perspective of Symbolic Convergence might be extended for use in studying this medium of social exchange in which identity and agency are constantly negotiated through strictly verbal means. If the promise of increased global communication is achieved, a deeper understanding of just how people make themselves understood as they cross national, cultural, and political boundaries is necessary. The chapter demonstrates that dynamic negotiations of identity and agency seem not to be confined to any single demographic stratum or subset of strata, and calls for the observation, description, and analysis of ongoing negotiations to grant us important clues about trans-national communication in the future.

Finally, Robert Shuter's "Ethical Issues in Global Communication" indicates that with the explosion of electronic messages worldwide, ethical issues have surfaced concerning the use of global communication technology. In this chapter the author addresses generic ethical issues that affect all communication technologies from telecommunication to the Internet. Based on the comparison of Western communication ethics and Islamic foundations of ethical communication, the author proposes that an intracultural approach is an alternative to a universal ethics in global communication. The intracultural ethical perspective possesses four characteristics: (1) It is based on the assumption that ethical communication standards are wedded to culture; (2) It relies on grounded analysis to do ethical critiques rather than universal ethical frameworks; (3) It challenges communication ethicists to immerse themselves intellectually and emotionally in the culture of the community being studied; and (4) It requires communication ethicists to explore and explain possible effects of their personal backgrounds on their ethical analysis.

Part III deals with the nature and forms of interaction in global society. In addition to Babara Monfils' end remarks, five chapters are included in this part. D. Ray Heisey's "Global Communication and Human Understanding" raises the question: How do world leaders attempt to bring about better understanding of conflicting ideologies and cultures in the globalizing society? The author addresses this question by examining two recent events that are major intersections of world history between the East and West. The first is the international visit to the U.S. by China's President Jiang Zemin when he was seen by America close-up and America was seen close-up by the Chinese leader and his people. The second is an international conversation with Iran's President Khatami where Iran, now isolated, was heard around the world and the U.S. was seen being given a chance to respond to an historic gesture. Because of global communication the leaders involved in these rhetorical interactions can be seen facilitating increased human understanding of differing cultures and ideologies. These ideological contrasts are given human faces and voices in each of these intersections as the discourse texts of the leaders are examined in the chapter for their rhetorical visions of what the respective nations involved stand for. The human dimensions, as opposed to abstract national entities and stereotypes, are there mediated by global communication that in turn assists in creating an improved global equilibrium of human understanding.

Mary Jane Collier's "Reconstructing Cultural Diversity in Global Relationships: Negotiating the Borderlands" is an argument and set of analytical possibilities that is under construction, and that may be a useful model to guide our study of and our conduct in global relationships. The main tenets of the argument are that communication scholars need to conduct inquiry into cultural diversity by being critically reflexive, broadening our theoretical assumptions, and translating what we know across multiple perspectives. The author proposes that we can engage others (respondents and researchers) in dialogue, create a process of reconstructing and co-constituting our analytical frames, recognize the relationships between macro structures and micro practices, and ultimately enact intergroup and interpersonal relationships in the borderlands, the both/and "third" spaces and moments in time.

Sue Wildermuth's "Social Exhange in Global Space" indicates that as more and more people from all over the world join the on-line community, additional information about how on-line relationships form, survive, and decline is vital to the expansion of intercultural and interpersonal communication research. In this chapter the author examines on-line interaction through the lens of social exchange theories. Findings reveal that the on-line context itself is neither inherently restrictive nor non-restrictive of interpersonal relationships and thus, of cultural diversity. Rather, it is individual perceptions that determine whether or not one chooses CMC for relationship formation. Based on perceptions of the context, individuals determine the costs and rewards of on-line interaction, and those cost/reward ratios, in turn, determine behaviors. Such findings carry extensive implications for understanding interpersonal and intercultural communication via Internet in the global society.

Leda Cook's "Conflict, Globalization, and Communication" discusses several popular frameworks for conflict analysis and intervention and puts the application of those models in the global context. Interpersonal, intercultural, and international models for the analysis of conflictive communication are compared and assessed for their effectiveness in describing global scenes. To accomplish this task, conflict and globalization are described as they are used in the chapter; then, using Levinger and Rubin's (1995) framework, interpersonal and international contexts are compared according to their different impacts on the conflict process. Next, several frameworks for analyzing intercultural conflict and the explanatory potential of such models for accounting for globalization processes are discussed. Finally, a model of conflict analysis that emerges from a particular global "scene" is then proposed.

Finally, William J. Starosta and Guo-Ming Chen's "Listening across Diversity in Global Society" examines global listening. The paradigm shift from the study of message transfer and reproduction to the examination of third culture building that transpires between interactants lays the basis for the examination in this chapter. A model of global listening is developed. The model delineates four elements of global listening: (1) the economics of listening, (2) listening within Culture One versus Culture Two, (3) barriers to global listening, and (4) double-emic listening. The authors argue that listening among those of differing cultures moves from an intracultural stage (Culture One) to an interpersonal stage (Culture Two). It then enters an intercultural stage of relationship building and restructuring, and a rhetorical stage of internalizing any changes. Finally, it enters a new intracultural stage at a more global level. The rhetorical stage is one in which variant values are negotiated and adjusted among parties so as to render them mutually palatable. The restructured view of the world that follows global listening can be termed Culture Three. The authors further indicate that research is required on listening across diversity in global society to see how the processes of presentation, interaction, adjustment, and internalization can function more smoothly. Movement from Culture One Through Culture Two to Culture Three should be carefully researched and documented to learn how global listening can promote diversity, mutual learning, and peaceful interaction among nations and peoples.

CONCLUSION

The trend toward globalization may be arrested, but not denied within human society. The revolution of technology has been gradually rupturing the wall between nations and between cultures. The approach of 21st century will signify the fading of traditional society and bring forth a brand-new psychological pattern and life style. If individuals or nations are incapable of riding the wave of globalization by developing a new pattern of thinking and behavior, they will soon be swallowed by the trend. Because communication and transportation technology are the main impetus that stimulate globalization and intercultural communication, new knowledge and skills are required for people to survive in the 21st century. This chapter proposes a research outline communication scholars can follow in order to generate necessary principles that can help people adapt to the coming global society.

Three research categories proposed in this chapter, namely, continuity and change of identity in global community, emergence and impact of global media, and nature and forms of interaction in global society, are believed to be the critical issues human beings must face in the future society, and are also vital to a broader range of communication research. With the collaborative efforts to tackle problems embedded in these areas, we can foresee that a better future for human society will be achieved.

REFERENCES

Adler, N. J. (1997). *International dimension of organizational behaviors*. Cincinnati, OH: South-Western.

Adler, P. S. (1982). Beyond cultural identity: Reflections on cultural and multicultural man. In L. A. Samovar and R. E. Porter (Eds.), *Intercultural communication: A reader* (pp. 389–405). Belmont, CA: Wadworth.

Belay, G. (1993). Toward a paradigm shift for intercultural and international communication: New research directions. In S. A. Deetz (Ed.), *Communication Yearbook, 16* (pp. 437–457). Newbury Park, CA: Sage.

Boulding, E. (1988). *Building a global civic culture*. New York: Teachers College.

Chen, G. M. (1998). Intercultural communication via e-mail debate. *The Edge: The E-Journal of Intercultural Relations, 1, 4*. Retrieved November 15, 1998 from the World Wide Web: http://kumo.swcp.com/biz/theedge/chen.htm.

Chen, G. M., & Starosta, W. J. (1996). Intercultural communication competence: A synthesis. *Communication Yearbook, 19*, 353–384.

——. (1998). *Foundations of intercultural communication.* Boston, MA: Allen & Bacon.

Collier, M. J., & Thomas, M. (1988). Cultural identity: An interpretive perspective. In Y. Y. Kim and W. B. Gudykunst (Eds.), *Theories in intercultural communication* (pp. 99–120). Newbury Park, CA: Sage.

Cooper, M. M., & Selfe, C. L. (1990). Computer conferences and learning: Authority, resistance, and internally persuasive discourse. *College English, 52*, 847–869.

December, J. (1996). Units of analysis for Internet communication. *Journal of Communication, 46*, 14–38.

Giddens, A. (1991). *Modernity and self-identity*. Stanford, CA: Stanford University Press.

Featherstone, M. (1990). *Global culture*. Newbury Park, CA: Sage.

——. (1991). *Consumer culture & postmodernism*. Thousand Oaks, CA: Sage.

Frederick, H. H. (1993). *Global communication and international relations*. Belmont, CA: Wadsworth.

Friedman, J. (1994). *Cultural identity and global process*. Thousand Oaks, CA: Sage.

Garnsey, R., & Garton, A. (1992, September). *Pactok: Asia Pacific electro-media gets earthed.* Paper presented via the Adult Open Learning Information Network Conference, Australia.

Harasim, L. M. (1993). Global networks: An introduction. In L. M. Harasim (Ed.), *Global networks: Computers and international communication* (pp. 143–151). Cambridge, MA: The MIT Press.

Hart, W. B. (1998). Intercultural computed-mediated comunication. *The Edge: The E-Journal of Intercultural Relations, 1, 4.* Retrieved November 15, 1998 from the World Wide Web: http://kumo.swcp.com/biz/theedge/iccmc.htm.

Herman, E. S., & McChesney, R. W. (1997). *The global media.* London: Cassell.

Kim, S. (1998). Cultural imperialism on the Internet. *The Edge: The E-Journal of Intercultural Relations, 1, 4.* Retrieved November 15, 1998 from the World Wide Web: http://kumo.swcp.com/biz/theedge/kim.htm.

Lynch, J. (1992). *Education for citizenship in a multicultural society.* London: Cassell.

Ma, R. (1994). Computer-mediated conversations as a new dimension of intercultural communication between East Asian and North American college students. In S. Herring (Ed.), *Computer-mediated communication.* Amsterdam, Netherlands: John Benjamins.

Morgan, G. (1988). *Riding the waves of change: Developing managerial competencies for a turbulent world.* London: Jossey-Bass.

Naisbitt, J. (1994). *Global paradox.* New York: Aven.

Peck, M. S. (1992). The true meaning of community. In W. B. Gudykunst & Y. Y. Kim (Eds.), *Readings on communication with strangers* (pp. 435–444). New York: McGraw.

Porter, R. E., & Samovar, L. A. (1994). An introduction to intercultural communication. In L. A. Samovar and R. E. Porter (Eds.), *Intercultural communication: A reader* (pp. 26–42). Belmont, CA: Wadsworth.

Rhinesmith, S. H. (1993). *A manager's guide to globalization.* Chicago, IL: Irwin.

Richmond, A. H. (1994). *Global apartheid: Refugees, racism, and new world order.* New York: Oxford University Press.

Robertson, R. (1992). *Globalization: Social theory and global culture.* Newbury Park, CA: Sage.

——. (1995). Glocalization: Time-space and homogeneity-heterogeneity. In M. Featherstone, S. Lash, R. Robertson (eds.), *Global modernities* (pp. 25–44). Thousand Oaks, CA: Sage.

Stevenson, R. L. (1994). *Global communication in the twenty-first century.* New York: Longman.
Thayer, L. (1987). *On communication.* Norwood, NJ: Ablex.

PART I
Continuity and Change of Identity in Global Community

CHAPTER TWO
Dialectics of Globalization and Localization

Rueyling Chuang
St. John's University

> There is always a dialectic, a continuous dialectic, between the local and the global.
>
> — Stuart Hall

The emerging globalization propelled by new technology and mass media in the 80s and 90s exemplifies the thrust of postmodernism. In the postmodern era, the boundary between global versus local becomes blurry. Featherstone (1990) contends that postmodernism is both a symptom and image of the global culture enriched by diversity, a variety of local discourses, codes, and practices. In a similar vein, Hannerz (1990) contends that the world culture is "marked by an organization of diversity rather than by a replication of uniformity" (p. 237). On the other hand, globalization also becomes a catalyst for the emergence of third cultures and a unified world culture. Thus, the dialectics between globali-zation and localization and the ramifications of an emerging global culture warrant our attention. The extent to which a unified world culture affects cultural diversity, and the maintenance of local identity should be carefully examined.

The dialectics between global and local culture can be exemplified by Hannerz's (1990) observation: the world culture is created through the increasing interconnection of varied local cultures. Appadurai (1990) posits that the central problem of globalization is the dialectic tension between cultural homogenization and heterogenization. However, Dezalay (1990) argues that "globalization is for the most part an Americanization" (p. 281). The Americanization of a global culture via technology and mass media might lead to political hegemony, cultural imperialism, and intellectual colonization, and consequently diminish local identity.

This chapter seeks to explicate the dialectic contradictions between localization and globalization. The author examines the implications of a unified world culture as the consequence of globalization, as well as the homogenization and heterogenization of local cultures. The foci of the paper include the dialectical process of cultural integration, cultural disintegration, and the effects of an emerging transnational culture, global culture, and a third culture.

CHARACTERISTICS OF GLOBALIZATION PROCESSES

To envisage the dialectics of globalization and localization, a conceptual framework undergirding the characteristics of globalization and its impact on local society and cultural identity is warranted. The concept of globalization and its pertinent terms such as global culture, global identity, global field and global human condition have drawn attentions of numerous theorists in the 90s (Featherstone, 1990, 1991; Friedman, 1990; Robertson, 1991, 1992; Wallerstein, 1990). Globalization is perceived as a frame of reference and an emerging paradigm. Featherstone and Lash (1995) are in sync with this observation that in the early 90s globalization has become an increasingly influential paradigm in the research. Robertson (1992) explains that globalization is not simply a matter of societies, regions, and cultural life being compressed together in different problematic ways, but also increases intensity inside nationally constituted societies. Globalization has affected the dynamics of nation-state, local and national, the formation of *Gemeinshaft* (community), and *Gesellshaft* (society). Robertson (1991) states that the ideas of culture, globalization, world-system, and identity have become increasingly problematic. Under the influence of the recent globalization with its manifestation of global economy, global mass media and advertising, global cultural life, and global hybridization, local and national identity have become increasingly fuzzier and more problematic.

As a result, the relativity of old and new identities and old and new ethnicities emerged due to global migration and interethnic interaction. The global field, or global human condition, involves processes of relativization that indicate ways in which "globalization proceeds, challenges are increasingly presented to the stability of particular perspectives on, and collective and individual participation in, the overall globalization process" (Robertson, 1992, p. 29). In other words, the relativization of citizenship, relativization of self-identities, relativization of societies, relativization of societal reference constitute the essential parts of this global field in which the distinction between individual and society and politics and humanity becomes problematic due to its complexity and relativity.

Robertson (1992) points out that although the term "global identity" has been used in contradictory ways, it has been a part of "global consciousness" and centered upon "global" (p. 8). The concern with globality and globalization cannot be deemed merely as an outcome or byproduct of Western modernity, because globalization is multidimensional and involves comparative interaction of different forms of life. This argument is in contrast to Giddens' (1990) contention that the consequence of modernity leads to globalization.

Regardless of the disagreements among scholars, globalization and global identity undergo an historical transformation. The term "globalization" has been developed recently (Featherstone, 1990; Robertson, 1992). Mazlish (1997) postulates the conceptualization of the global epoch. He mentions that the globalization process in recent years transcends existing local, regional, and national boundaries. For example, recent environmental concerns and telecommunications such as satellites indicate an emerging global identity. Coincidentally, in the emerging global world the self-consciousness of civilization, society, ethnicity, regions, and individuality is also accentuated (Robertson, 1992).

In reaction to Robertson's argument that globalization cannot be primarily seen as a product of Westernization or modernity, Pieterse (1995) claims that "the modernity/globalization view is not geographically narrow (Westernization) but historically shallow (1500 plus)" (p. 47). Pieterse's contention illustrates the multiplicity of globalization, and that it does not primarily involve universalization but rather interculturalism (cf., Featherstone & Lash, 1995). With the hybridization of global culture, we have a new combination of new cultural practice. What we have is cultural *mélange,* a mixture of various cultural forms, a combination of global and local, such as Chinese tacos and Asian rap in London (Pieterse, 1995). Globalization is then not a condition of modernization, but a historical epoch starting from the 1960s that is associated with postmodernity. Modernity, however, is contemporaneous with an earlier period (1840–1960) which belongs to the "hegemony of the nation-state" (Featherstone & Lash, 1995, p. 5). Further, conceptualization of globalization as a product of Westernization, Western standardization, and cultural synchronization (e.g., Schiller, 1989) is fundamentally incomplete and too simplistic. As Pieterse (1995) notes, cultural synchronization overlooks the counter-currents—the impact of non-Western cultures on the West. It also downplays the ambiguity of the "globalizing momentum and ignores the role of local reception of Western culture" (p. 53).

The assumption that globalization is primarily Westernization, Europeanization, and Americanization is limited and problematic. In fact, globalization could be Japanization or Brazilianization (cf., Featherstone & Lash, 1995). For example, the Japanization of the global economy and market cannot be overlooked. The pervasive consumption of Japanese products (e.g., cars and electronic appliances) in American and Chinese households is a clear indication of the Easternization/ Japanization of the global economy. Thus, the true globalization is hybridization that is also associated with postmodernism (Pieterse, 1995; Featherstone, 1990). In this postmodernity, the diverse voices (i.e., global and local) are valued. The dialectic of globalization and localization is an ongoing process in which the center and peripheral/marginal have been taken into account.

MODERNITY, POSTMODERNITY, UNITY, AND DIVERSITY

Along with the debate over whether modernity leads to globalization, the extent to which globalization ties into modernity or postmodernity, unity, and diversity warrants this chapter's attention. Robertson (1992) acknowledges that individuals, movements, and institutions have not only participated in actions that propelled the burgeoning globalization process, but simultaneously have been resistant to this. The resistance and acceptance of globalization on the local level have created a force of dialectical contradiction. Robertson (1992) further notes that the term globalism has a negative connotation, which has been associated with ideological terms as one-worldism and cosmopolitanism. Although negative gestures (i.e., gestures of opposition) are often displayed in "contemporary terms and in reference to contemporary circumstances" (p. 10), in spite of a certain denial of global wholeness, the trends towards the unicity of the world are inexorable and necessary. Robertson uses the economic protectionism in the early 1990s as an example of the negative responses to increasing compression of the unified world. In addition, the new protectionisms are "more self-consciously situated within a globewide system of global rules and regulations concerning economic trade and a consciousness of the global economy as a whole" (p. 26). It is thus understandable that globalization can potentially threaten the economic well being of a local culture, and on a larger scale, the existence of a local culture. The dialectical tension between the local and the global becomes a constant challenge to the globalization process. How a local culture adopts the global culture and becomes competitive in the global arena, while sustaining its own cultural identity and sovereignty, becomes a central part of the dialectical process.

Several theorists share the argument that globalization leads to unified and essentially Westernized world culture and is the consequence of modernity. For example, Hall (1991b) contends that the new forms of globalization are related to global mass culture, which remain centered in the West. Global mass culture deals with, according to Hall, technology, advanced labor, capital, and its form of homogenization.

Through the technology, commercial, and cultural synchronization originated from Western society, it is evident that globalization is closely linked to modernity. Other theorists also argue that the emergence of global culture or globalization inevitably leads to a unified world culture or world culture system in which stronger cultures (e.g., Western and American cultures) devour weaker cultures (e.g., local and tribal cultures). Dezalay (1990) remarks that globalization of economy entails the globalization of the market relating to finance, production, commerce, and industry. He uses North American law firms and mega-law firms as examples to indicate that deregulation and globalization of the market for legal services was essentially an Americanization.

While assuming globalization as essentially an Americanization phenomenon (e.g., CocaColonization, McDonaldization, Hollywoodization), one might also notice that Japanese culture globalizes Asian countries. The consumption of Japanese products such as Sony CD players and Nintendo game gear indicates the globalization of Japanese economy. Along with Dezalay's argument, one can take Taiwan for an example: the deregulation of Taiwanese industry, such as the tobacco and liquor industries, leads to Americanization of the Taiwanese market. The pervasiveness of Japanese culture and commercial products in other Asian countries is another manifestation of the global Japanization.

Tenbruck (1990) as well considers that the developing cultures, such as the independent nations of Africa, have absorbed European *modus operandi* (e.g, science, education, and language). The liberation of colonial countries, together with the velocity of modernization, has increased their dependence on developed countries. These developing countries look toward countries such as the United States of America and the Soviet Union that represent their interests on a global level. The superpower nations are the center of cultural prestige that form the transmission belt for the cultural development of developing countries. This inevitably leads to Americanization and Russification of the respective "spheres of influence" of the developing nations (p. 204). Tenbruck concludes that the gradual colonization of European cultures (also Western culture or American culture) would be absorbed. Today, English has advanced to the status of *lingua franca*. It has become a common (and probably the only) language which allows diverse ethnic groups of Asians, Indians, or Africans to communicate with each other. Even among homogenous cultures such as Chinese, given the fact there are a wide variety of dialects, a Hong Kong Chinese person and a Taiwanese Chinese would probably need to rely on English to carry on a lengthy conversation. In this scenario, global language such as English has become a medium to bridge the language differences among local people.

In addition to English or Russian as a *lingua franca* of the cultural elite in their particular sphere of influence, Tenbruck (1990) also notes that both languages have become the linguistic medium for the sciences and everyday language. The pervasiveness and global presence of electronic mass media, which rely on English as their primary *lingua franca*, has accentuated the importance of English. English has become more than ever a world language and a signifier for world culture. Tenbruck further remarks that the global development will continue to form and dissolute people, languages, cultures, nations, states, and local identity. The expansion of European cultures and American culture has led to "ubiquitous presence and ongoing interpenetration of all cultures" (p. 204). These global development has accelerated global migrations and dislocation of people and consequently propelled the local cultures into globalwide efforts of presenting their ideas and lifestyles overseas. These multi-cultural conjunctures (or according to Appadurai's term, disjunctures) will affect and perhaps threaten the survival, preservation, dissolution, and extinction of (local) cultures. American television, Tenbruck observes, habitually diffuses American history, cultural patterns, and the way of life. This leads individual cultures to lose their autonomy. For example, the diffusion and popularity of American TV sitcoms, films, music, and fashion have affected the youth in various cultures. People in different countries consume American pop culture willingly, follow the American trends closely, and idolize American pop culture artists. In this sense, the dialectic of globalization and localization also links to the evolution and dissolution of cultural identity.

In addition to the constant dialectical tension of globalization and localization, there are two views regarding whether globalization creates a unified culture that perpetuates cultural domination and cultural imperialism. Some worry that cultural domination and cultural imperialism will penetrate a global ideology and pose a threat to the maintenance of local identity and sovereignty. Robertson (1992) points out that the capitalism and imperialism have forced the world into a compressed condition. Widespread Western capitalism and imperalism have crystallized the potential problem of a contemporary global economy. In spite of Schiller's (1993) reservation that media-cultural imperialism is only a subset of the general system of imperialism, he argues that the trend toward globalization may ameliorate the increasing economic and social disparities around the world.

Schiller (1993) cautions that the "internationalism based on either the weakness or the opportunism of most of the participants can hardly be viewed as a movement toward global equilibrium and social peace" (p. 107). The actual sources of globalization are not to be found in a newly achieved harmony of interests in the international arena. The infrastructure of what is seen as the internationalism should be sustained by the transnational corporate business order and in all forms of economic, consumption, production, and distribution. Schiller argues that corporate media-cultural industries (e.g., Time Warner) occupy global social space. Cultural domination is not limited to the exposure to American television shows. Cultural submersion includes the English language itself, fashions, theme parks (e.g., Disney), American pop culture, watching CNN news, and dining at American fast-food restaurants (hence McDonaldization). Though American cultural domination is not perpetual or detrimental to local economy, undeniably American cultural domination has been preeminent for the past forty years and continues to be so (Schiller, 1993). With the popularity of electronic media such as the World Wide Web and Internet, one might argue that the globalization of American culture would diffuse other local cultures on a deeper level and on a larger scale. It is apparent that with the new technology, which is dominated by American industry (e.g., IBM, Intel & Microsoft), to keep up with the global economy, a local culture or nation would have to learn to adapt. In this sense, the American culture can be deemed the embodiment of global culture. The colonization of American technology and industry might erode the local economy and thus potentially create a cultural imperialism.

Featherstone (1991) envisages the connection between globalization and postmodernity as seeking to promote diversity rather than universal wholeness or cultural imperialism. By advocating the connection between globalization and postmodernism, Featherstone expresses the dialectical concern of homogeneity and heterogeneity in relation to globalization, localization, and postmodernity. He argues that, though the emerging globalization process coincides with the view that the world is one, global culture does not lead to homogeneity or a unified culture, but rather a widening range of different cultural contacts with others. In other words, it is not valid to state that globalization creates the tendency to cultural integration and ultimately homogenization by assuming that, for example, multinational capitalism, Americanization, Coca Colonization, media imperialism, and consumer culture will diminish local differences. According to Featherstone (1990), postmodernism is both a symptom and a cultural image of the conceptualization of global culture. Postmodernism and global culture seek to sway away from the so-called homogenizing processes, cultural imperialism, Americanization, or Westernization of economical, political, and cultural domination. Instead, postmodernism seeks to promote diversity that includes a variety of local discourses, codes, and practices, but resists systemicity and order. To Featherstone (1991), the concept of postmodern culture is derived from Western context with its emphasis on polyculturalism and integration of "otherness" of different cultural traditions. It seeks to produce inclusive and "pluralistic global circumstances with some tendencies towards cultural disorder" (p. 123).

THIRD CULTURE AND WORLD CULTURE

In addition to the characteristics of globalization that closely tie into relativization of the global human condition (as postulated by Robertson), postmodernity (as advocated by Featherstone) and the hybridization of global culture (as proposed by Pieterse), the emergence of world culture and third culture merits our further consideration. Hannerz (1990) points out that the so-called world culture is "marked by an organization of diversity rather than by a replication of uniformity" (p. 237). Hannerz's assertion is similar to the conception that globalization is related to postmodernity and emphasizes diversity. With the integration and interpenetration of various cultures, it is quite possible that a new form of the so-called "third culture" is created.

Featherstone (1991) observes that the burgeoning flow of finance, immigrants, images, and information has created "third cultures" that are transnational and mediate between national cultures. They go beyond the level of inter-state exchanges. Yet, Featherstone notes, there is a further sense in which we can talk about a global culture: the process of global compression whereby the world becomes united to the extent that it is regarded as one place (see also, Robertson, 1990).

Featherstone (1990) argues that a global culture can be both global integration and disintegration taking place not only on the inter-state level but also on a trans-national or trans-societal level. These trans-societal global processes include the exchange and flow of products, people, information, knowledge, and visual images that give rise to communication processes on a global level. Consequently, third cultures, which include a variety of diverse cultural flows, emerge. Featherstone therefore argues that it is not accurate to perceive that a global culture will weaken the sovereignty of nation-states. The speculation that the nation-states will be absorbed into a unified world, which produces cultural homogeneity and integration, is misleading. Moreover, the emergence of a third culture is not the embodiment of homogenization. The polarized logic which offers explanation of a global culture via the essentialized terms such as homogeneity/heterogeneity, integration/disintegration, unity/diversity, must be abandoned (Featherstone, 1990).

GLOBALIZATION AND LOCALIZATION IN FLUX

The fluid dialectical interaction between localization and globalization is best represented by Hall's (1991a) conception of local and global. Hall notes that we ought to consider the globalization processes and the dialectics of local/global as contradictory formulations. Globalization has accelerated the erosion of the nation-state, national economies, and national cultural identities. However, the new forms of globalization simultaneously occur in two opposite directions: "it goes above the nation-state and it goes below it. It goes global and local in the same moment." Global and local create "two faces of the same movement from one epoch of globalization, the one which has been dominated by the nation-state, the national economies, the national culture identities, to something new" (p. 27). This new national culture is the result of the integration of global and local culture. To Hall, the global mass culture and the "American conception of the world" are interwoven (see also King, 1991).

Hall also indicates that the new kind of globalization deals with a new form of global mass culture which is dominated by the image crossing and re-crossing linguistic frontiers rapidly by television, film, and mass advertising. Hall uses satellite television to illustrate that global mass media as such cannot be constrained by national boundaries. Global mass culture has two characteristics (Hall, 1991a): First, it remains in the West. The driving force of the global mass culture lies in Western technology, the concentration of capital and techniques, and the visual imagery of Western societies (e.g., films, TV shows). However, the global mass culture speaks English as an international language. Though the language (e.g., English) is centered in the West, it is no longer centered in the same way. It is not standardized as that of highbrow English. The local adaptation and variation of a global language such as English can be seen by the examples of American English, African American English, Caribbean English, and Bahamian English. This is a representation of dialectical interaction between globalization and localization: the global shapes the local and is simultaneously reshaped by the local.

Hall (1991b) further argues that globalization is pertinent to a hegemonic sweep at which "a certain configuration of local particularities try to dominate the whole scene, to mobilize the technology, and to incorporate, in subaltern positions, a variety of more localized identities to construct the next historical project" (p. 67). There is always a dialectical interaction between local and global, the question to Hall is the locations where struggles might evolve, global or local. Hall continues to argue that counter-politics positioned at the level of confronting the global forces that are trying to recapture the world are not making much progress. However, local places seem to have the ability to develop counter-politics, resistance, counter-movements, and countercurrents. The places where they emerge as a counter-political force are localized even if they are not necessarily local.

Moreover, the global is encompassing, creating similarity, working through particularity, and negotiating particular spaces, ethnicities, and identities (Hall, 1991b). A dialectical contradiction always exists between local and global and the two are constantly interpreting each other. Hall's conception of the dialectical interaction of global and local holds true in many countries. Again, using Taiwan as an example, it is apparent that Taiwan is both bombarded by the constant need and urge to become more global in terms of economy, technology, and industry, and by the emerging need of preserving its tribal and local cultures. On one hand, Taiwan seeks to become more competitive in the globe; on the other hand, the need to sustain its local culture and cultural value has become ever more essential.

Other issues intertwining with the dialectics of global and local are the concerns regarding nation/state, center/peripheral, inside/outside, *Gemeinschaft* (community)/*Gesellschaft* (society), and universalism/particularism. As Robertson (1992) notes, the relationship between the universal and the particular is essential to our understanding of the ramifications of globalization processes. The universalistic ethics and morals have been rejected by many poststructuralists and postmodernists (e.g., Featherstone, 1991). Robertson (1992) not only contends that the ideas of culture, globalization, world-system, and identity are problematic, but also argues that in a world that has been increasingly compressed while its inter-state systems are subject to internal and external constraints, the conditions of and for the identification of individual and collective selves/others have become ever more complex. Indeed, the conception and boundary of the global/local, state/nation, universalism/particularism, Gemeinshaft/Gesellshaft, and cosmopolitan/local is increasingly blurry, fragmented, and multidimensional. This, as previously mentioned, can be deemed the manifestation of postmodernity.

GLOBAL-LOCAL NEXUS

In addition to the dialectical interaction, the concept of global-local nexus provides a different explanation of the recent cultural phenomenon. Robertson (1992) suggests that rather than simply viewing universalism as a principle applicable to all (i.e., global) and particularism to local, the global and local should be tied together as part of "globalwide nexus" (p. 102). The two are connected based on the universality of the experience and the expectation of particularity. The "particularization of universalism" involves the notion of the universal being given global-human concreteness, while universalization of particularism includes the widespread idea that there is no boundary to particularity, "to uniqueness, to difference, and to otherness" (Robertson, 1992, p. 102).

Appadurai (1990) states that the essential problem of recent global interactions is the dialectical tension between cultural homogenization (i.e., globalization) and heterogenization (i.e., localization). A related debate about homogenization is the inter-linked argument about Americanization and commoditization. The fears of Americanization can be extended to different nations, such as the Korean's fear of Japanization and the Baltic Republic's fear of Russianization. Appaduari clearly points out that for "polities of smaller scale, there is always a fear of cultural absorption by polities of larger scale, especially those that are near by" (p. 295). Globalization's scalar dynamic ties closely with the dialectical relationship between nations/ states, globalism/tribalism, and local/cosmopolitan. Thus, the new global cultural economy should be comprehended as a complex, overlapping, disjunctive order. The simple model of push and pull (a form of dialectical contradiction), the existing center-periphery model, or the dialectics of consumers and producers alone cannot offer a complete picture of the global cultural economy. Appaduari insists that even the most complex theories of global development grounded in Marxist tradition (also Wallerstein, 1974) are inadequate in explaining the global cultural economy phenomenon.

Since theorists such as Appadurai find the dialectical opposition of the global/local and center/peripheral models inadequate, the conceptualization of glocalization provides an alternative view of the recent globalization phenomenon. Glocalization captures the dynamics of the local in the global and the global in the local. It essentially illustrates the global localization condition. Robertson (1995) depicts glocalization as a blending of global and local. Citing the definition from *The Oxford Dictionary of New Words*, Robertson notes that glocalization has been "modelled on Japanese *dochakuka* (deriving from *dochaku*, living on one's own land), originally the agricultural principle of adapting one's farming techniques to local conditions, but also adopted in Japanese business for global localization, a global outlook adapted to local conditions" (p. 28). For Robertson there is no such dialectical tension in glocalization. He uses the concept of glocalization to point out that the bipolar view of global/local is problematic, because it is inaccurate to assume that the local is always against global trends. The two can co-exist harmoniously. In other words, local/global and *Gemeinshaft/ Gesellshaft* are not always at opposite ends and local is not always the counterforce of global. Local should not be seen as an "interpretative departure point, as a counterpoint to the global" (Robertson, 1995, p. 30).

Accordingly, globalization does not necessarily homogenize or standardize locality. Robertson agrees with Abu-Lughod's (1994) assertion that globalization has involved the reconstruction and reintegration of the so-called "home," "community," and "locality." Coincidentally, Pieterse's (1995) argument that globalization should be seen as a hybridization phenomenon and Hannerz's (1990) argument that "there can be no cosmopolitans without locals" (p. 250) are in accord with Robertson's (1995) concept of glocalization. The tandem operation of global/local dynamics and glocalization, according to Pieterse (1995), can be related to examples such as minorities who seek support for local demands from transnational channels. As he indicates, globalization means the integration of localism, as "Think globally, act locally" (p. 49).

CONCLUSION

This chapter seeks to synthesize and bring to light the different views on the ramifications of globalization in the post/modern era. The globalization phenomenon is closely tied to global culture, global economy, world culture, transnational culture, and third culture. The conceptualizations of globalization as an historical epoch, a frame of reference, a paradigm, and the global field/global human condition pertinent to the relativization of societies and citizenship are discussed. Theorists' various concerns with regard to globalization as essentially Westernization, Europeanization or Americanization, which might lead to cultural imperialism, are delineated. Accordingly, globalization might hegemonize and colonize local and national identity. Media conglomerates, American technology, and commodification (i.e., Coca Colonization, McDonaldization, Hollywoodization of the entertainment industry) are manifestations of the compressing globalization to the point that it might bring forth a unified world.

However, other theorists (e.g., Featherstone & Lash, 1995; Pieterse, 1995; Robertson, 1995) hold that globalization is a representation of postmodernity in which marginal voices are privileged and diversity is valued. Globalization is considered to entail the ongoing integration and interpenetration of locality and globality and to be multidimensional and complex. Consequently, to perceive globalization based on the bipolar and static model of state/nation, heterogeneity/homogeneity, convergence/divergence, local/global, local/cosmopolitan, and *Gemeinshaft/ Gesellshaft* is problematic. The complexity of globalization cannot simply be explained by the assumption that the aforementioned distinctions are mutually exclusive or in constant opposition. As Hall (1991b) conceptualizes, global and local should be seen as a fluid dialectical interaction. This is not to say that local should always be seen as the counterforce of global, and the *Gemeinshaft* (community) is in constantly resistance to the *Gesellshaft* (society). In addition to this, the dynamics of globalization can be seen as hybridization (i.e., a mixture of global and local) or as glocalizaton and global-local nexus in which the local is in the global and the global is in the local (i.e., global localization).

REFERENCES

Abu-Lughod, J. (1994). Diversity, democracy, and self-determination in an urban neighborhood: The East Village of Manhattan. *Social Research, 61* (1), 181–203.

Appadurai, A. (1990). Disjuncture and difference in the global cultural economy. In M. Featherstone (ed.), *Global culture: Nationalism, globalization and modernity* (pp. 295–310). Newbury Park, CA: Sage.

Dezalay, Y. (1990). The Big Bang and the law: The internationalization and restructuration of the legal field. In M. Featherstone (ed.), *Global culture: Nationalism, globalization and modernity* (pp. 279–294). Newbury Park, CA: Sage.

Featherstone, M. (1990). Global culture: An introduction. In M. Featherstone (ed.), *Global culture: Nationalism, globalization and modernity* (pp. 1–14). Newbury Park, CA: Sage.

——. (1991). *Consumer culture & postmodernism.* Thousand Oaks, CA: Sage.

Featherstone, M. & Lash, S. (1995). Globalization, modernity and the spatialization of social theory: An introduction. In M. Featherstone,

S. Lash, and R. Robertson (eds.), *Global modernities* (pp. 1–24). Thousand Oaks, CA: Sage.

Friedman, J. (1990). Being in the world: Globalization and localization. In M. Featherstone (ed.), *Global culture: Nationalism, globalization and modernity* (pp. 311–328). Newbury Park, CA: Sage.

Giddens, A. (1990). *The consequences of modernity*. Stanford, CA: Standford University Press.

Hall, S. (1991a). The local and global: Globalization and ethnicity. In A. King (ed.), *Culture, globalization and world-system* (pp. 19–40). Binghamton, NY: State University of New York at Binghamton.

——. (1991b). Old and new identities, old and new ethnicities. In A. King (ed.), *Culture, globalization and world-system* (pp. 41–68). Binghamton, NY: State University of New York at Binghamton.

Hannerz, U. (1990). Cosmopolitans and locals in world culture. In M. Featherstone (ed.), *Global culture: Nationalism, globalization and modernity* (pp. 237–252). Newbury Park, CA: Sage.

King, A. (1991). Introduction: Spaces of culture, spaces of knowledge. In A. King (ed.), *Culture, globalization and world-system* (pp. 1–18). Binghamton, NY: State University of New York at Binghamton.

Mazlish, B. (1997). Psychohistory and the question of global identity. *The Psychohistory Review, 25* (2), 165–176.

Pieterse, J. N. (1995). Globalization as hybridization. In M. Featherstone, S. Lash, R. Robertson (eds.), *Global modernities* (pp. 45–68). Thousand Oaks, CA: Sage.

Robertson, R. (1990). Mapping the global condition: Globalization as the central concept. In Featherstone (Ed.), *Global culture: Nationalism, globalization and modernity* (pp. 15–30). Newbury Park, CA: Sage.

——. (1991). Social theory, cultural relativity and the problem of globality. In A. D. King (ed.) *Culture, globalization and the world-system* (pp. 69–90). Binghamton, NY: State University of New York at Binghamton.

——. (1992). *Globalization: Social theory and global culture*. Newbury Park, CA: Sage.

——. (1995). Glocalization: Time-space and homogeneity-heterogeneity. In M. Featherstone, S. Lash, R. Robertson (eds.), *Global modernities* (pp. 25–44). Thousand Oaks, CA: Sage.

Schiller, H. I. (1989). *Culture Inc*. New York: Oxford University Press.

——. (1993). Not yet the postimperialist era. In C. Roach (Ed.), Communication and culture in war and peace (pp. 97–116). Newbury Park, CA: Sage.

Tenbruck, F. H. (1990). The dream of a secular ecumene: The meaning and limits of policies of development. In M. Featherstone (ed.), *Global culture: Nationalism, globalization and modernity* (pp. 193–206). Newbury Park, CA: Sage.

Wallerstein, I. (1974). *The modern world system.* New York: Academic Press.

——. (1990). Culture as ideological battlegroud of the modern world-system. In M. Featherstone (ed.), *Global culture: Nationalism, globalization and modernity* (pp. 31–56). Newbury Park, CA: Sage.

CHAPTER THREE
Dialectics of Identity and Diversity in the Global Society

Mei Zhong
San Diego State University

Moving toward diversity is a trend in most human societies in recent years. Diversity has begun to remodel the way we live and has emerged as a norm rather than an exception in modern world. Psychologically, we feel the world becoming smaller and find ourselves living in a "global village." Because of this, changes are inevitable in many people's lives. People may choose to deal with this condition actively or passively, but we all recognize that we must face the issue in order to succeed and live productively in the 21st century. This leads more and more people to face a great dilemma on a daily basis: how to maintain their unique individual identities while striving to fit into a diverse society. This chapter examines this particular issue from a communication perspective. More specifically, this chapter will (1) state the trend toward diversity, with specific reference to the situation in the United States, (2) survey and examine the literature on identity by focusing on identity formation at both individual and social group levels and its connection to communication, and (3) investigate the dialectical relation between identity and diversity.

DIVERSITY TREND IN THE GLOBAL SOCIETY

Many scholars have been interested in investigating why people of different societies are becoming more interdependent in the modern world. For example, Gergen (1991) identified seven technologies of social saturation that account for this trend: railroad, mail, automobile, telephone, radio, motion pictures, and commercial publishing. We can add television and computer, Internet in particular, to the list because both are revolutionary in terms of their influence on modern society and human communication behaviors. As Gergen argued, "Each of these technologies brought people into increasingly close proximity, exposed them to an increasing range of others, and fostered a range of relationships that could never have occurred before" (p. 53).

Martin and Nakayama (1997) summarized several reasons to study intercultural communication that reflect why human societies become more diverse and why effective intercultural communication is vital for us to live in the future world. Among them, three are especially relevant to the discussion of a diverse society: fast-advancing technologies, ever-faster immigration patterns, and interdependent economic styles. In addition, we see that the change in international political climate, namely the shift of attention from military to domestic affairs as a result of the end of the "cold war era," has its role in promoting human interdependence and demanding intercultural communication competency. The tremendous change in international political climate has led to the use of advanced technology in areas other than military, thus allowing people to travel and communicate much more freely and easily. The development of the Internet, for example, makes it possible for people to "meet" and interact with others in all kinds of places without leaving home.

The end of the cold war also promoted free traveling and migration. More countries are open for travel to the rest of the world and the U.S., as a superpower in the world, represents a country of prosperity. The increase of immigration along with the existing domestic movement provides people of differing cultural and ethnic backgrounds an opportunity to live and work together. More people are assigned to work in foreign countries, or engaged in international, interethnic, and interracial relationships. People also find themselves living in an environment drastically different from the one they used to know. As Hwang (1994) stated, ". . . we see the rise of a new demographic reality: a nation with no majority race" (p. xi).

Particularly, Juan (1994) pointed out that the Asian American population, the third largest minority group in the U.S., has doubled during the last decade. It has even been projected that UCLA will become an Asian American campus by the year 2000. However, the size of population does not equal diversity in power. The conscious and systematic struggle for equal rights can be dated back to the civil rights movement. Hwang (1994) asserted that "Certainly whites continue to control a wildly disproportionate amount of power in the United States; that injustice has not changed" (p. xi). While the beginning of the civil rights movement was primarily for promoting African-American cultures, an awakening of the need for a stronger sense of self and identity among other groups is emerging. More cultural groups are now actively engaging in the mainstream political process.

Another result of the change in international politics is that the world reflects more interdependence among countries. It is not unusual to see a single product made with efforts across several nations. Meanwhile, more companies are setting up branches outside their base countries in order to compete in international markets. With the end of the "cold war," the focus of world politics is changing from military and nuclear contests to domestic economic development. This transformation of international politics may have been the determining factor for most of the changes and development in other areas. It may have indirectly caused a faster pace toward a more diverse world. Increasingly people are engaged in situations where diversity becomes a part of their normal life.

Politics also played an important role in diversifying the community in certain countries. Even in presumably homogeneous societies such as the People's Republic of China, the mixture of people with different cultural backgrounds is increasing. The end of the "Mao era" saw an open-door policy and political reform that brought about tremendous changes in China. For centuries, China had been a closed society to the rest of the world. The communist ruling in the past half a century, in particular, has been known as sealing China with an "iron curtain." However, China's outlook has been drastically changed over the past two decades. For example, economic development has brought numerous joint ventures between China and the rest of the world. As a result, many foreigners are working in China and many Chinese are working in these joint ventures across major cities in China. Another example is that educational exchange programs with international institutions have brought in foreign experts and foreign cultures to China. These changes have resulted in a more heterogeneous society within China.

The diversification of the United States of America and People's Republic of China represents and brings about more frequent interactions within and among other countries in the world. As cultures and races are getting more interdependent, people tend to embrace and learn from each other on one hand, while feeling anxious with and frustrated by problems created by diversity on the other hand. The phenomenon reflects one of the problems in a diversified society: the need to keep one's identity. We not only need to learn to appreciate diversity but also need to look into ourselves or our own group for insights and strength in order to successfully handle our daily situations. It is conceivable that by interacting with other cultures, we come to a stronger sense of the ourselves. The dialectical relationship between diversity and identity warrants the further examination of the concept of identity. As Dunbar

(1997) stated, "There is an increasing interest in the assessment of social group identity and collective self-esteem" (p. 1).

THE ISSUES OF IDENTITY

Two main issues of identity have been examined by scholars in communication discipline: the construction or formation of identities and the relationship between identity and communication.

Identity Formation

Identity formation can be approached from two basic levels: "I" and "we." According to Tanno and Gonzalez (1998), the "I" is identity at the individual level and the "we" is group identity. Research that examines identity at the individual level focuses on the psychological construction of the concept as well as the influence of social interaction. This line of research studies identity in relation to self-concept, self-esteem, and self-image. In psychologist Erikson's term, it is the idea of "ego." Erikson (1956) maintained that ego identity is within the self and is the "choices" and "decisions" one makes for the self. These "choices" and "decisions" arrive at a final definition of self with a sense of commitment. In contrast, Slugoski and Ginsburg (1989) argued that ". . . it [identity] is predicated upon an impoverished and highly delimited conception of society, and we demonstrate that its ultimate reliance on internal ego-integrative processes results in a normative model of identity that is class-, race-, and sex-bound" (p. 37). Moreover, Mead's idea of the "self" also serves to explain how we construct our individual identity through symbolic interaction (Blumer, 1969; Mead, 1934, 1962). Mead (1934) indicated that the self is something to be developed. Self does not exist at birth, rather, it "arises in the process of social experience and activity" and "develops in the given individual as a result of his [sic] relations to that process as a whole and to other individuals within that process" (p. 135). Mead (1934) identified two stages that account for the process of fully developing the self: (1) the individual organizes the attitudes of other individuals toward him/herself and others during social interaction, and (2) the individual organizes the social attitudes of the generalized other, which is usually a collective attitude of the whole community or social group.

A more relevant level of identity in regard to diversity is the social and group identity. Recent research stresses the need for the individual as well as the cultural group to maintain a healthy sense of cultural identity and to ensure minority empowerment (Cho, 1994; Juan, 1994; Kim, 1994). Scholars have agreed with two general ways of identity formation.

For example, Tajfel and Turner (1979) found that social identity is based on the ascription to ingroup attitudes and the formulation of attributions about out-group individuals. Similarly, Martin and Nakayama (1997) described avowal and ascription as the two processes in group identity formation. According to the authors, avowal is the self-portrayed identity and ascription is the other-attributed identity. In addition, Dunbar (1997) identified three ways that a group identity is formed: (1) ascribed in-group identity, (2) self-empowerment via group membership, and (3) perceived social attribution and support assigned to group membership. The ascribed in-group identity is "the active choice to refer to one's self as a member of a designated social ingroup" (p. 2). The self-empowerment via group membership is when "the personal sense of control, self-efficacy, and affective attachment derived from belonging to a specified social group" (pp. 2–3). And the perceived social attribution is when "social group identity incorporates the acknowledgement of how group membership is perceived in the large social environment" (p. 3).

Most discussions relating identity to global society focus on cultural group identity. According to Collier (1988), "Cultural identity emerges in everyday discourse and in social practices, rituals, norms, and myths that are handed down to new members" (p. 131). Martin and Nakayama (1997) specifically pointed out that cultural identity is influenced largely by history. Dunbar (1997), on the other hand, pointed out seven categories of identity: age, ethnic heritage, gender, race, religion, sexual orientation, and socioeconomic status. This is similar to Martin and Nakayama's (1997) six categories of one's social and cultural identities: gender, age, race and ethnic, national, regional, and personal. Others may argue that additional categories, such as profession and political orientation, may also be considered categories of social and cultural identity.

As for specific research on identity and culture, Chen (1992) examined Chinese American women's identity as portrayed by the media with stereotypical images. As a Chinese-American herself, Chen (1992) stated that "I often find myself struggling between two sets of cultural realities, each with its own beauty and coherence, as well as contradictions and limitations" (p. 226). This demonstrates a typical dilemma of co-cultural living and calls for attention to the conflict between diversified societies and the struggle of individuals with their cultural identities.

Other studies focus on specific symbolic meanings of identity. For example, Lee (1998) looked at foot bonding as a part of traditional Chinese women's identity by examining how the action and result of foot bonding symbolizes beauty and, therefore, gives Chinese women their identity in society. Pratt (1998) reported the use of humor as identification among American Indians by looking at how patterns of humorous language determine the tribe of the user. Gergen (1991) explored the identity of the self through language accent of the speaker. Roland (1988) studied specific cultural identities of the Asian Indian and Japanese cultures by looking at the familial self, the individual self, and the spiritual self in these cultures. And Hegde (1998) studied Asian Indian immigrant women's identity in the U.S. by concluding that identity is ongoing negotiations between subjective experience and external representations faced in the practices of daily life.

Furthermore, Kirkpatrick (1985) noted that although identity construction/formation is a psychological process, it may be possible to recognize that "people of different cultures construct their worlds differently without proposing either an extreme particularism or broad generalizations that reduce human insights about persons and social process to West versus the rest, we versus they formulas" (p. 238). In doing so, "some of those insights may become available not only as examples of the human imagination but also as contributions to a more profound understanding of the interrelations of self and social life" (p. 238). In light of such recognition, Martin and Nakayama (1997) proposed specific stages of identity development for majorities and minorities in today's societies, especially in the United States. According to them, there are five stages in the development of a majority identity: (1) The unexamined identity, (2) the acceptance stage, (3) the resistance stage, (4) the redefinition stage, and (5) the integration stage. The unexamined identity is when one is unaware of the implications of the self and what the self stands for. The acceptance stage is the internalization, conscious or unconscious, of a racist ideology that could be passive or active. The resistance stage is one in which members move passively or actively from blaming the minority to blaming their own dominant group as the source of racial problems. The redefinition stage is when members refocus, redirect, and redefine "whiteness" (this is in reference to the majority in the U.S.) in nonracist terms. Finally, the integration stage is when members of the majority group recognize "whiteness" as self after redefinition and accept others as a result.

Likewise, there are four stages in the minority identity development. The unexamined identity signifies one's own original ethnicity. The conformity stage is the internalization of values and norms of the group. The resistance and separatism stage refers to the period of dissonance or a growing awareness that not all dominant group values are beneficial to minorities. Endorsement of one's own group values and rejection of those of the dominant group often happens at this stage. Finally, the integration stage gives members of the minority group a strong sense of self and their own group identity, and an appreciation of other cultural groups as well (Martin & Nakayama, 1997).

Identity and Communication

How is identity expressed? What is the role of communication in identity construction? Martin and Nakayama (1997) indicated that "Issues of identity are particularly important in intercultural communication" (p. 64). According to them, we learn who we are through communication and then communicate who we are to others. Identity is expressed through language and labels and is "complex and subject to negotiation" (p. 89). Gumperz and Cook-Gumperz (1982) also linked identity to communication by stating that the spoken language can be simultaneously examined from three perspectives: (1) language usage, which is the actual verbal message, (2) inferencing, the interpretation of the language, and (3) evaluation, which determines "how participants reflexively address the social activity that is being constituted by their ongoing talk" (p. 19). These perspectives refer to how we use, infer, and evaluate meanings with people of our own culture, and how we do these differently in and with other cultures.

Saldivar (1990) studied Chicano narrative as a way of understanding the Mexican-American culture and stated the important "vital" connection between historical genesis of an ethnic minority and its narrative. The author maintained that " . . . basically, . . . reality can be known in experience only if it is first *imagined* as a formed product of the subjects who recognize it or misrecognize it and express it in symbolic forms" (p. 212). Saldivar's point is demonstrated through his emphasis of using autobiography as a symbolic form in constructing the Chicano identity. He argued that "because of its fundamental tie to themes of self and history, self and place, it is not surprising that autobiography is the form that stories of emergent racial, ethnic, and gender consciousness have often taken in the United States and elsewhere" (p. 154). Because of their viable importance, autobiographies and their discourses may be used to "chart the historical, cultural, and psychological factors surrounding the development of an ideology of the self" (p. 154).

Martin and Nakayama (1997) have pointed out that sometimes misperceptions of one's identity may create problems when we interact only according to stereotypes. We may choose to use, infer, and evaluate language differently from what is intended. This obviously should be incorporated into our study of the identity issue. They pointed out three approaches to the understanding of identity: (1) social psychological perspective, (2) communication perspective, and (3) critical perspective.

The social psychological perspective emphasizes that "identity is created in part by the self and in part in relation to group membership" (p. 64). This perspective recognizes that "the self is composed of multiple identities, and that these notions of identity are culture bound" (p. 64). The communication perspective, on the other hand, emphasizes that "identities are not created by the self alone but are co-created through communication with others, . . . identities emerge when messages are exchanged between persons" (p. 67). This idea is similar to Hecht, Collier, and Ribeau's (1993) assertion that identities are negotiated, co-created, reinforced, and challenged through communication, and the specific vehicles are the avowal and ascription processes. As to the critical perspective, it attempts to understand identity formation within the contexts of history, economics, politics, and discourse.

DIALECTICS OF IDENTITY AND DIVERSITY

Some researchers have explored the relationship between diversity and identity or unity. For example, Louw (1998) examined the issue of "cultural economy" in South Africa and raised the question of whether stability is more important than development. Others argued for the importance of maintaining identity of cultural groups (e.g., Cho, 1994; Juan, 1994; Kim, 1994). However, the key question is how can we function as effective members in diverse society and maintain a strong and healthy cultural identity at the same time? Three aspects of intercultural communication studies, including theory, research, and real life, can be used to answer this question.

At the theory level, intercultural adaptation theory is a good example for demonstrating the tension between identity and diversity, i.e., how can we find the ideal balance and keep a leveled state on the continuum of our intercultural adaptation? Basically, we need a greater acculturation level to successfully adapt to a host culture; but at the same time, we also need a good sense of identity to find peace within ourselves. Individuals without a healthy sense of cultural or group identity usually tend to lose the ability to relate to others in the process of adaptation. According to Kim (1988), when we try to adapt, we move toward a higher degree of diversity awareness and often give up part of our identity in the process of "deculturation." If we insist on maintaining our original identity, an obstacle toward successful adaptation could possibly appear. What, then, is the optimal point for a person on the adaptation continuum? Is there a universal range for such a state? Is it just an individual or situational phenomenon?

A person's perceived satisfaction of intercultural adaptation is often based on expectation. For example, a Chinese graduate student rated himself as close to 100% adapted in the U.S. culture after a year and a half living here. He explained that he came here with one expectation and one expectation only: to study and to study well. For the year and a half he had been here, he was able to study in his chosen major, receive advice on his program, and get a good grade. To him, his goals were met and he was fully satisfied (Zhong, 1996). However, he experienced embarrassments in social life because he could not function well due to language barriers. Upon this realization, he quickly rated his adaptation level down to 70%, 60%, and even 40% (Zhong, 1996). So what should be considered as a successful adaptation? How much do we need to feel that we are of our own nationality, and how "Americanized" (in the U.S.) need we fill in the process of adaptation? Is it more important to keep in touch with one's original cultural identity by limiting interaction with others/hosts, or to dive into the host culture and try to become one of its members? These questions can be illustrated by the hesitation of international students in the U.S. to decide whether to use their native language or English in social interactions. For instance, a Chinese student trying to speak only English even with his fellow Chinese was often regarded as being Americanized and frowned upon by his Chinese friends. Reversely, when a group of Chinese converse in their mother tongue in public settings, they could be considered rude.

People caught between cultures often have to ask themselves questions about diversity and identity and determine a comfortable position on the continuum. It is important to determine a comfortable point, because it is the basis of successful adaptation and healthy self-esteem. As long as we live in a culturally diverse society like the United States of America, we are bound to constantly negotiate our own newly defined identities. According to Kim (1988), in many situations we try to fit into the diverse society, but we are unable to put aside our own cultural or group identity. In order to tackle this problem, the ability of behavioral flexibility, as a component of intercultural communication competence, is recommendable (Chen & Starosta, 1996). Behavioral flexibility refers to the ability to switch back and forth between or among cultures.

At the research level, the study of intercultural communication competence can be used to illustrate the dialectical relationship between identity and diversity (Chen, 1990, 1992). Research has shown that the traditional Western-based measures are not necessarily accurate in measuring competence in different cultures (Chen, 1993; Hedge, 1993; Miyahara, 1993; Yum, 1993). For example, verbal skills are essential for communication competence in Western cultures whereas many eastern cultures discourage such traits as oral expressiveness. The problem highlights the tension between identity and diversity: Should individual measures be developed to acknowledge unique cultural preferences or should a universal measure that reflects diverse cultural orientations be developed? In other words, do we need to account for a person's cultural values and develop scales that measure his/her skills accordingly or should we simply try to be inclusive and measure all possible dimensions in a scale? Moreover, knowing that we could not identify an individual measure across all cultures, do we choose to focus on certain specific cultural groups or shall we assume that the world will become more uniform, and therefore there will be a measure that fits the so called "general global culture"?

The real life level is the last example for explicating the tension between identity and diversity. The issue concerns the official language used in diversified cultures. For example, a few years ago, certain areas of southern California demanded that Spanish be established as the official language for the region because the majority of residents there are Hispanic descendants. In this situation, should the law be made in order to promote cultural identity by accepting the demand? If so, would this encourage other co-cultural groups to demand their own native languages to be used as official languages? The English vs. French dispute in Quebec is another example. Thus, how important is it to have a unified language? Would a unified language help to reinforce unity and destroy identity? All these questions show that the language we speak serves as an important tool for us to construct reality and establish our cultural identity. But we often have to compromise our identity to achieve more satisfactory communication with people in a culturally diverse society. The dilemma created by the dialectical tension between identity and diversity will continue to exist as the trend is moving to a global world. It will also demand scholars to explore the problem in order to search for a way to balance the pulling and pushing between the two forces.

CONCLUSION

So are we fighting a losing battle? Is there a perfect solution to maintaining a balance between identity and diversity? Is "multiple identity" reachable? Should we promote more diversity by emphasizing a unified society with diverse cultures, as in a "melting pot"? Should we focus more on our identities in the hope maintaining our own unique and colorful selves, as in a "salad bowl"? Although this chapter attempts to investigate the dialectical relation between diversity and identity, more questions rather than answers are raised.

The globalization is a trend that cannot be reversed. As we are inventing programs to educate our new generations to become genuine global citizens, we have to find the key that promotes diversity while enabling cultures and individuals to maintain their unique identities. Learning cultural uniqueness and promoting co-existence is the only way for us to live productively in the 21st century.

REFERENCES

Blumer, H. (1969). Symbolic interactionism: Perspective and method. Englewood Cliffs, NJ: Prentice Hall.

Chen, G. (1990). Intercultural communication competence: Some perspectives of research. *Howard Journal of Communications, 2*, 243–261.

——. (1992). A test of intercultural communication competence. *Intercultural Communication Studies, 2*, 63–82.

——. (1993, November). *A Chinese perspective of communication competence*. Paper presented at the annual meeting of the Speech Communication Association, Miami, FL.

Chen, G. M., & Starosta, W. J. (1996). Intercultural communication competence: A synthesis. *Communication Yearbook, 19* (pp. 353–383). Newsbury Park, CA: Sage.

Chen, V. (1992). The construction of Chinese American women's identity. In L. F. Rakow (Ed.), *Women making meaning: New feminist directions in communication* (pp. 225–243). New York: Routledge.

Cho, M. Y. (1994). Overcoming our legacy as cheap labor, scabs, and model minorities: Asian activists fight for community empowerment. In K. A. Juan, D. H. Hwang, & M. A. Jaimes (Eds.), *The state of Asian America: Activism and resistance in the 1990s* (pp. 253–273). Boston, MA: South End Press.

Collier, M. (1998). A comparison of conversations among and between domestic culture groups: How intra- and intercultural competencies vary. *Communication Quarterly, 36,* 122–144.

Dunbar, E. (1997). The personal dimensions of difference scale: Measuring multi-group identity with four ethnic groups. *International Journal or Intercultural Relations, 21,* 1–28.

Erikson, E. H. (1956). The problem of ego identity. *Journal of the American Psychoanalytic Association, 4,* 56–121.

Gergen, K. J. (1991). *The saturated self: Dilemmas of identity in contemporary life.* New York: Basic Books.

Gumperz, J. J., & Cook-Gumperz, J. (1982). Introduction: Language and the communication of social identity. In J. J. Gumperz (Ed.), *Language and social identity* (pp. 1–21). London: Cambridge University Press.

Hecht, M. L., Collier, M. J., & Ribeau. S. (Eds.). (1993). *African American communication.* Newbury Park, CA: Sage.

Hegde, R. S. (November 1993). *Passages from tradition: Communication competence and gender in India.* Paper presented at the annual meeting of the Speech Communication Association, Miami, FL.

——. (1998). Swinging the trapeze: The negotiation of identity among Asian Indian immigrant women in the United States. In V. T. Tanno & A. Gonzalez (Eds.), *Communication and identity across cultures,* (pp. 34–55). Thousand Oaks, CA: Sage.

Hwang, D. H. (1994). Facing the mirror. In K. A. Juan, D. H. Hwang, & M. A. Jaimes (Eds.), *The state of Asian America: Activism and resistance in the 1990s* (pp. ix-xii). Boston, MA: South End Press.

Juan, K. A. (1994). Linking the issues: From identity to activism. In K. A. Juan, D. H. Hwang, & M. A. Jaimes (Eds.), *The state of Asian America: Activism and resistance in the 1990s* (pp. 1–15). Boston, MA: South End Press.

Kim, E. H. (1994). Between black and white: An interview with Bong Hwan Kim. In K. A. Juan, D. H. Hwang, & M. A. Jaimes (Eds.), *The state of Asian America: Activism and resistance in the 1990s* (pp. 71–100). Boston, MA: South End Press.

Kim, Y. Y. (1988). *Communication and cross cultural adaptation.* Philadelphia: Multilingual Matters.

Kirkpatrick, J. (1985). How personal differences can make a difference. In K. J. Gergen & K. E. Davis (Eds.), *The social construction of the person* (pp. 225–240). New York: Springer-Verlag.

Lee, W. (1998). Patriotic breeders of colonized converts: A postcolonial feminist approach to antifootbinding discourse in China. In V. T. Tanno & A. Gonzalez (Eds.) *Communication and identity across cultures* (pp. 11–33). Thousand Oaks, CA: Sage.

Louw, E. (1998). "Diversity" versus "national unity": The struggle between moderns, premoderns, and postmoderns in contemporary South Africa. In V. T. Tanno & A. Gonzalez (Eds.), *Communication and identity across cultures,* (pp. 148–174). Thousand Oaks, CA: Sage.

Martin, J. N., & Nakayama, T. K. (1997). *Intercultural communication in contexts.* Mountain View, CA: Mayfield.

Mead, G. H. (1934). *Mind, self, and society*. Chicago: University of Chicago Press.

——. (1962). *Mind, self and society*. Chicago: University of Chicago Press.

Miyahara, A. (November 1993). *Communication competence: A Japanese perspective*. Paper presented at the annual meeting of the Speech Communication Association, Miami, FL.

Pratt, S. B. (1998). Razzing: Titualized uses of humor as a form of identification among American Indians. In V. T. Tanno & A. Gonzalez (Eds.) *Communication and identity across cultures* (pp. 56–79). Thousand Oaks, CA: Sage.

Roland, A. (1988). *In search of self in India and Japan: Toward a cross-cultural psychology*. Princeton, NJ: Princeton University Press.

Saldivar, R. (1990). *Chicano narrative*. Madison, WI: The University of Wisconsin Press.

Slugoski, B. R., & Ginsburg, G. P. (1989). Ego identity and explanatory speech. In J. Shotter & K. J. Gergen (Eds.), *Texts of identity* (pp. 36–55). London: Sage.

Tajfel, H., & Turner, J.C. (1979). An integrative theory of intergroup conflict. In W. G. Austin & S. Worchel (Eds.), *The social psychology of intergroup relations* (pp. 33–48). Monterey, CA: Brooks-Cole.

Tanno, D. V., & Gonzalez, A. (1998). Sites of identity in communication and culture. In V. T. Tanno & A. Gonzalez (Eds.), *Communication and identity across cultures*, (pp. 3–7). Thousand Oaks, CA: Sage.

Yum, K. (1993, November). *Communication competence: A Korean perspective*. Paper presented at the annual meeting of the Speech Communication Association, Miami, FL.

Zhong, M. (1996, November). *Chinese students and scholars in the U.S.: An intercultural adaptation process*. Paper presented at the annual meeting of the Speech Communication Association, San Diego, CA.

CHAPTER FOUR
Reconfiguring the Global Society: "Greater China" as Emerging Community

Hui-Ching Chang
University of Illinois at Chicago

INTRODUCTION

The increasing contact and mutual interdependence among people of different nations, particularly in light of modern technological advancement, have led many scholars to observe that a global society is in the making. Just by clicking a mouse, for example, one can get in touch with others in remote locations, unfamiliar societies, and distant cultures. In such a global society, members derive their sense of identity not only by belonging to one specific culture or nation, but by simultaneous participation in multiple societies.

People are no longer constrained to their locale or immediate physical context but now are able to reach out to the global society. As Clark (1997) notes, "The information revolution has greatly increased the number of people who experience the world as a single place, but more important, it has transformed the dimensions, indeed the very nature, of that experience" (p. 146). We now experience the world in real time and on demand in an interactive mode, seemingly without even being aware of such an experience (Clark, 1997).

Because of this ability to get in touch with people of different cultures/nations, the dialectic relation between the local and the global is increasingly contested, since now one simultaneously belongs to one specific nation/culture/society and to the "global village" as well. The struggle between the localized sense of identity and the need to reach out and assume multiple levels of existence in the modern world becomes unavoidable (Chen & Starosta, 1998; Naisbitt, 1994). The global society, thus conceived, is a massive web wherein people constantly construct, re-construct, and co-construct the meaning(s) of their existence through their seemingly contradictory attempts to keep to the status quo in order to maintain their identity while at the same time extending themselves into multiple domains of experience.

However, while greater global interdependence and technological advancement allow people in different parts of the world to orchestrate and inter-connect their lives into new patterns, colors, and shades, it also, as Clark (1997) notes, ". . . is bringing about heightened cultural self-awareness and thus greater ethnic, religious, racial, and linguistic conflict" (p. 16).

The new configuration of a global community depends not only on technological advancement, but also on international political and economic concerns. The vacuum left by the end of the Cold War has challenged the "citizens" of the global society to re-structure a new world order, and consequently has problematized the rigidity of traditional national boundaries. This challenge is particularly acute, given the increasing mobility among people of different racial/ethnic groups, and the improved communication efficiency brought about by the information superhighway.

Following the end of the Cold War, with its bipolar assignment of nations either to the "democracy" or "Communist" camp—that is, aligned either with America or the Soviet Union—Mowlana (1996) notes the emergence of two seemingly opposing trends. On the one hand, there was an increased nationalism, manifested in national and ethnic conflicts in the former Yugoslavia, Czechoslovakia, and Soviet Union, as well as in resistance by some European nations to participation in the European Community. On the other hand, however, one could also observe a resurgent universalism. As Mowlana (1996) explains,

> . . . the end of the Cold War has removed the ideological divisions that divided different social groups. Most obviously the reunification of the two Germany's symbolizes this national reunion. At a larger level, separated communities in the East and the West can now associate with greater ease and less stigma. Religious, linguistic, and national groups can now assert the common bonds previously divided by the superpower conflict. In essence the Berlin Wall was a cultural as well as an ideological barrier. (p. 202)

There are other indicators of this universalistic movement. We have seen the European Union's plan to align all European nations as one economic entity, as well as the desire of a large Muslim population in new Central Asian states to create a single Islamic community (Mowlana, 1996). Finally, and still a situation very much in need of resolution, is the idea of community in what has come to be called "Greater China."

The struggle between nationalism and universalism reflects an urge to jettison outmoded nation-state boundaries and to redraw boundaries and form new alliances. As Chang (1995) mentions, some observers ". . . assert that the nation-state is receding in importance and is being increasingly marginalized by other political formulations . . . including supranational cultural-ethnic entities variously characterized as civilizations, economic origins, and global tribes" (p. 955). Politically, the sense of community can be established by various means of demarcation—whether by political systems and ideological stance (Democracy versus Communism), religion (Christianity versus Islam), geographic location (European Union versus North American Free Trade Association), or cultural/ethnic heritage (Chinese versus non-Chinese communities).

Kotkin (1993) also argues that the configuration of international economy and the diminishing importance of the ideology that once divided superpowers during the Cold War has given rise to what he calls the "global tribe." According to this idea, rather than following traditional national boundaries, people are demarcated as "tribes" who share common origin and established business and cultural networks worldwide, even though they may be dispersed geographically (Chang, 1995). While in the past such a situation would not be conceivable (due primarily to the constraints of physical location) modern technology has made it possible for such alliances to thrive.

The purpose of this chapter is to analyze the possibility of forming a Chinese community constructed by various Chinese political entities such as China (the People's Republic of China, or PRC), Hong Kong, and Taiwan (the Republic of China, or ROC), currently expressed by the concept of "Greater China." Specifically, this chapter will discuss the tension between Taiwan and China with respect to their visions of a unified China and the implications of how such unity may redefine our notion of a "global society." Although Kotkin's (1993) argument refers primarily to economic situations, the questions of whether various Chinese groups can transcend their nation-state boundaries to form a "Chinese tribe," based upon shared cultural foundations and civilization, for the purpose of extending their political and economic power (Chang, 1995), remain to be explored.

With approximately one-quarter of the world's population Chinese, an emergent community of Chinese nation-states would doubtless have a significant impact on the evolution of the global system. Is the formation of Chinese community a necessary historical trend, or are there forces of international power and politics, beyond the control of the parties involved, working to eventuate such an outcome? Is the urge to become "one glorious community" an unrealistic dream of some Chinese, or are there only practical and obvious political benefits involved? On what grounds can the two political entities negotiate? Will it be possible for China and Taiwan to go beyond their national boundaries and form some kind of alliance, either between themselves or in league with other Chinese countries? Perhaps most fundamentally, how does the discourse relating to the possibility of a unified "Chinese tribe," or even simply the possibility of cooperation among various Chinese groups, serve to change the balance of power in the international community? It is time we consider whether the formation of a Chinese community would create an alternative frame against which meanings of a global society can be interpreted.

THE CASE OF THE CHINESE COMMUNITY: THE CALL FOR UNIFICATION

To understand the possible construction of a new Chinese community, it is essential to discuss the relation between Taiwan and China. According to Harding (1994), "Taiwan is an instance of both of the processes—foreign colonialism and civil war—that created a divided China" (p. 236). Having long honored themselves as living in the culturally superior Middle/Central Country, citizens of "China" —whether the PRC or ROC—have, since the Qing dynasty, seen their countries fall prey to Western and Japanese imperialism.

China was divided into two political entities in 1949, after the end of Sino-Japan war (Wang, 1985). Since the Communist Party had claimed power on the mainland, the Kuoming Tang (KMT) in 1948 moved to Taiwan and formed the Republic of China, declaring the occupation of mainland China by the Communist Party illegal. The ROC had been considered to be the only "legitimate China" in the world until 1971, when China replaced Taiwan in the United Nations General Assembly and on the Security Council, and thereby became in the eyes of the world the only "real" China.

The problem of the "two Chinas" has long been an unresolved issue for scholars of international politics and law (Chen & Reisman, 1972). Since China established formal diplomatic ties with the United States in 1978, it has called on Taiwan for peaceful unification between the two nations. Responding to these overtures, the ROC has consistently reiterated that it considers China an "illegal government," codifying in

1981 its "three-no's" policy—no contacts, no negotiation, and no compromises with the PRC (Chao & Myers, 1998).

The PRC has tried to maintain its political and economic competitiveness in the world by gradually opening its doors to other nations, a trend begun by the implementation of an "open door" policy in the late 1970s. After the collapse of the Soviet Union and Eastern Europe, the PRC experimented with marketization by allowing free markets in specified areas such as southern China, and has, step-by-step, loosened the totalitarian or authoritarian governance endorsed by Communist ideology. Taiwan, on the other hand, after living under the rulership of Japan for the first half of the twentieth century, has gone on to become one of the most powerful economic entities in the world. Taiwan's degree of democratization is particularly impressive, leading some observers to claim that Taiwan is the first free Chinese country ever to exist (Garver, 1997).

Although a mere decade ago, the PRC and the ROC were still trying to use military force to defend themselves from one another, in recent years there have been dramatic fluctuations in the relationship between the two countries (Morgan, 1993; Regnier, Niu, & Zhang, 1994). Despite their apparent differences in political rules, ideological bias, physical size, economic development, and other characteristics, after many years of separation China and Taiwan have begun to narrow their differences and restore some common cultural bonds (Lin & Robinson, 1994).

In recent years, China has opened its doors to the Taiwanese people, encouraging investment, and softening the tone of its calls for unification. Similarly, Taiwan has changed its "three no's" policy, and since 1986, has gradually opened its doors to China, by acts such as allowing Taiwanese citizens to visit China, and to allow China's scholars and others to visit Taiwan. In 1990, guidelines for the National Unification Council were established by President Lee to clearly express Taiwan's endorsement of the policy of unification and the conditions under which Taiwan would consider such a situation workable. It should be noted, however, that China strongly objects to any attempt to create an independent Taiwan, and has said it will not hesitate to use armed force against the Taiwanese people in the event of any move toward independence. As of now (1998), only twenty-six countries, most of them small, still maintain diplomatic ties with Taiwan.

In an effort to coordinate extensive economic and cultural exchanges between China and Taiwan, various non-official organizations have also been set up and some semi-governmental officers have exchanged visits (Chao & Myers, 1998; Harding, 1994). With increased contact between

China and Taiwan, exchanges between Taiwan and Hong Kong have also become more frequent. Since neither Taiwan nor China has approved direct communication or transportation between themselves, Hong Kong has often acted as mediator between the two political entities.

Since the takeover of Hong Kong in 1997 by the People's Republic of China, the dynamics of the interlinkages among the PRC, ROC, and Hong Kong have shifted, bringing to the forefront recurring questions about how Chinese unity (that is, the sense of a common heritage) may be achieved. Although both governments agree there should be only "one China" in the world, they differ with regard to the terms and conditions under which unification could be achieved.

Resolving the Contradiction Within

In order to create a sense of Chinese community and to stress the need for "one China," both China and Taiwan must dissolve the apparent dissimilarities between themselves. What is the base or bases upon which China and Taiwan could be unified? Can unification be achieved around the idea of a Chinese "motherland"? Of what consequence is it that people of both political entities are "ethnically Chinese"? Is "cultural heritage" the base upon which commonality can be built? Or, is unification to be the inevitable consequence of "historical trends"? As we unfold these ideas, it is well to keep in mind that both China and Taiwan use alternative rhetorical strategies to create the impression of having a "common heritage."

The Mainland Chinese Stance. In arguing for a "One China" policy, China frequently refers to the fact that Taiwan has always been part of China, and that immigrants to Taiwan were originally people from China. As Wang (1985) notes, "To Peking, the reunification of the motherland is the responsibility of all Chinese, an approach via Chinese nationalism" (p. 43). The PRC describes the unification between the two political entities as an "historical trend," implying that it cannot, and should not, be changed artificially (Garver, 1997, p. 13).

China's arguments for a unified China entail both political and philosophical dimensions. The PRC sees such unification as an inevitable "historical trend" that cannot be reversed. It is particularly interesting to note how China provides reasons to back up its arguments about this "historical trend." The fact that all Chinese are descendants of the Han people is thought to justify the call for unity. Pye (1985) points out that Chinese have a strong sense of racial identity, and that "awareness of the importance of one's own ancestry seems to have made the Chinese feel that they have a bond with all people of Chinese blood" (p. 185).

According to this line of reasoning, ethnically, racially, and culturally, people of China and Taiwan belong to one country; they not only share the same language and have the same ancestors, but also inherit the same cultural traditions. Even though China and Taiwan differ in their sociopolitical and economic systems, the demarcation between Communism and democracy is downplayed. Instead, as Garver (1997) points out,

> One hoary Chinese idea . . . was the proposition that all Chinese should be united under a single emperor, the Son of Heaven, ruling over all under Heaven (Tian Xia). This notion stood at the very center of China's traditional political thought . . . The division of Chinese-populated territory into several states was considered an abnormality, an abomination, to be overcome as quickly as possible. (p. 17)

This point is similar to one made by Pye (1985), who contends that the idea of "one China" helps to maintain a unitary political system in China, and ". . . has made the Chinese uneasy whenever their cultural world has been sundered by contending authorities" (p. 184). From China's point of view, unification between China and Taiwan is not merely a pragmatic concern that will benefit both political entities in the future, but is considered to be a "responsibility" of the Chinese people (Huang, 1992). Returning to one's "motherland," an idea which exercises political power as well as moral influence (Tu, 1994a), thus becomes a result of historical necessity which must be handled with care. As Yahuda (1993) notes,

> As seen by leaders of the PRC the importance of unification with Taiwan is a matter of the civil war, completing a nationalist tryst with Chinese history and finally triumphing over the demeaning effects of past humiliations at the hands of Western powers. (p. 698)

Moreover, the constant fluctuation between expanding and shrinking, and between disintegration and integration of the Chinese empire, also leads Chinese to expect a return to a center of Chinese power (Harding, 1993). As one common Chinese saying puts it, "In the world situation, unity will necessitate disintegration, and disintegration will necessitate unity" (a notion derived from the classic Chinese novel, *The Tale of the Three Kingdoms*). Chinese philosophical traditions surrounding the "mandate of Heaven" reflects the way (tao) of the universe, and thus provides the foundation upon which political discourse is presented.

Of course, the fact that Taiwan's economy is so strikingly successful provides yet another incentive for China to argue that Taiwan should be under Chinese sovereignty, particularly since Taiwan has, in times past, been considered as a land populated by barbarians and indeed abandoned to the control of several different nations throughout China's history. From China's point of view, unification is something that ought to be considered seriously by people of both China and Taiwan. Philosophically, China couches its intentions in the argument that humans must follow the course of the universe which demands a return to unification from the abnormal, unnatural state of disunity, while politically, it is undeniable that China stands to gain much from Taiwan's enormous economic success.

The Taiwanese Stance. Taiwan's perceptions of the idea of "one China" are, in some important ways, completely at odds with those of the PRC. Although Taiwan has long nurtured dreams of independence, it was not until the lifting of martial law in 1987 that Taiwanese began to openly challenge the idea that China and Taiwan would one day be unified. While agreeing that people of China and Taiwan both speak the same official language, and also that they come from the same cultural traditions, many Taiwanese claim to have established a unique Taiwanese culture of their own, one which is noticeably different from that of mainland China. As Garver (1997) notes, "The transformation of cross-Strait relations interacted with a discourse on collective identity among the people on Taiwan. Were they Chinese or Taiwanese? If they were Chinese, what exactly did this mean?" (p. 16; see also Tu, 1994b). Or, to pose another parallel question raised by Copper (1996): Is Taiwan a nation-state or simply a province of China?

Under political and military pressure from the PRC, Taiwan's government, as represented by KMT, endorses the "one China" policy, although under terms different from those postulated by the PRC (Chao & Myers, 1998). According to the KMT, people in Taiwan are ethnically Chinese, although their experiences and success in democracy make the culture distinctive (Garver, 1997). As Tu (1994b) puts it, "In response to the threat of the independence movement, the government deems it advantageous to underscore Taiwan's Chineseness, but the challenge from the mainland prompts it to acknowledge how far Taiwan has already departed from the Sinic world" (p. 10). Hence, Taiwan's view of common heritage with Chinese on the mainland has less to do with racial commonality than with envisioning a future state of integration.

In the guidelines for the National Unification Council, established in 1990 by President Lee, it was stated that unification between the two political entities could be achieved only under principles of political democracy, social pluralism, market economy, respect for the Taiwanese people, and so on (Wei, 1994). The Taiwanese government has expressed its willingness to negotiate with the PRC to achieve some kind of unity, so long as the PRC has adopted democracy, thereby moving in the same direction as the Taiwanese have since emerging from the control of the Japanese, following World War II.

Officially, the ROC government agrees with China's argument that all Chinese people should be unified under one political system, saying, in effect, "we are all Chinese." Some political groups in Taiwan take an even stronger stand; for example, the New Party, spun off in 1994 from people dissatisfied with President Lee's leadership of the KMT, openly agrees with China's view of ". . . the Chinese-speaking people on Taiwan as ethnically and racially Chinese, as tong bao (born from the same womb) with mainland Chinese" (Garver, 1997, pp. 19–20).

However, other political groups in Taiwan take different positions. The DPP, Democratic Progressive Party, a major party opposing KMT, contends that there is little commonality, either historically or culturally, between Taiwan and China, and hence no need for unification (Chao & Myers, 1998). From DPP's perspective, many of those who still envision a "one, great China," whether they are in Taiwan or on the Mainland, are still strongly attached to the notion of nationalism and the "mandate of Heaven."

Although the KMT, the New Party, and the DPP differ in the extent to which they want to be unified with the PRC, all three are against the "collective identity" notion articulated by the PRC under its "one China, two systems"[1] policy (see Garver, 1997). Most believe that this policy amounts to treating Taiwan as China's subordinate. The viewpoint of many Taiwanese, seeking unification with China is more a matter of practical concern for the survival of Taiwan, given its limited international political status, and is thus less a matter of "responsibility" and the "mandate of Heaven." One common argument adopted by Taiwanese to refute China's reasoning is the analog between Taiwan and the United States: if such a "historical trend" toward reunification exists, why does America not want to reunite with the "motherland" of Great Britain? There is, this argument goes, no reason why Taiwan cannot establish its own identity, apart from the PRC, whether people in Taiwan want to establish an independent country or form some kind of alliance on an equal footing with the PRC.

Sense of "Chinese-ness"

It seems quite clear that, despite the rhetoric of unification, the contradictions and incompatibility between China and Taiwan encompass multiple dimensions (Tu, 1994a). What would be a desirable state for people in both political entities? Should it be "one country, two systems," as advocated by the PRC, or should it be "one country, two governments," as advocated by KMT? Or, should it be two separate countries forming some kind of alliance? The fundamental question of why there should be a need for only "one China," whether based on politics or philosophy, still remains unresolved. To take a different cut at this question, I propose to take up the idea of the sense of "Chinese-ness," an idea surfeit with ethnic, political, and historical complexities.

Being Chinese means being at the center of human existence which, while surrounded by barbarians, is derived from a superior civilization with an impressive history that has evolved for thousands of years (D. Wu, 1994). The Chinese civilization is said, according to legend, to have originated from an ever-expanding core area, the Wei River Valley in North China, where Chinese people developed a long history and thus made their cultural identity distinctive. The Chinese culture also symbolizes a unique form of life different from that of the "barbarians" (i.e., those who are not Chinese). Indeed, "the expression of hua or huaxia, meaning Chinese, connotes culture or civilization" (Tu, 1994b, p. 3). The Han people were believed to be the race of China, principally on the strength of having assimilated other minority groups who were thought not to be able to resist the superior Han civilization (D. Wu, 1994).

The sense of "Chinese-ness" results from the orchestration of culture, civilization, common ancestry, and documentary records of human history. As Tu (1994a) explains,

> Surely a salient feature of being Chinese is to belong to a biological line traceable, as the legend goes, to the Yellow Emperor. This ethnic identification, mythologized in the idea of the "dragon's seed," evokes strong sentiments of originating from the same progenitor. Since China proper, the Central Country (Zhongkuo), is clearly identifiable on the map of the world, another salient feature of being Chinese is simply being born in the Divine Land (shenzhou), a poetic name for China. . . Chineseness entails common ancestry, homeland, mother tongue, and basic value orientation. (p. v)

D. Wu (1994) notes that two sentiments have been shared by those who see themselves as Chinese: Zhongguoren (people of the Central Country) and Zhonghua minzu (the Chinese race or the Chinese people) (pp. 149–150).

The term Zhongguoren "carries the connotation of modern patriotism or nationalism . . . a connectedness with the fate of China as a nation. Associated with this is a sense of fulfillment, of being the bearers of a cultural heritage handed down from their ancestors, of being essentially separate from non-Chinese" (D. Wu, 1994, p. 149). The term Zhongguoren was introduced by the late-19th century Chinese statesman Zhang Taiyan, who proposed Zhongguoren as the authoritative definition of the Chinese people. According to Zhang, the ancestors of the Han people were centered in North China and called themselves Hua Xia. Zhang and his supporters,

> . . . concluded that Hua, Xia, or Han could be used interchangeably to mean China the nation-state, Chinese the race (or tribe), and China the geographic location. . . Zhang Taiyan's vague and inclusive definition marks the beginning of a modern concept of Chinese national identity. (D. Wu, 1994, p. 150)

Similarly, the term Zhonghua minzu emerged at the turn of the 20th century during the early Republic of China, and was often used in nationalistic writings to warn Chinese about the dangers of Western invasion. The two terms became conflated when Chinese faced Western invasion, and thus came to represent an identity based upon cultural and historical fulfillment rather than the more modern conception of citizenship (D. Wu, 1994).

The meaning of "Chinese-ness" must be viewed in the context of China's turbulent modern history and the suffering of the Chinese people, exacerbated by Western imperialism, the Taiping Rebellion, the end of Qing dynasty, Japanese aggression, the struggle between Nationalists and Communists, and many other political events and circumstances. This history has led China "from the imagined universally recognized cultural center (the 'Central Country') to the humiliating status of backwardness in virtually all areas of human endeavor" (Tu, 1994a, p. vi). This reversal of fortune helps perpetuate and create a paradoxical complex associated with "being Chinese"—one views oneself simultaneously as both superior and inferior to people of other cultures/nations, and thus the feeling of "Chinese-ness" becomes more pronounced, prominent, and complicated.

Although any discussion of "Chinese-ness" must remain "a layered and contested discourse" (Tu, 1994a, p. viii), Weng (1995) claims that Chinese ". . . generally subscribe to the notion that all Chinese belong to one nation and that there should be only one sovereign state of China" (p. 370). To have a single China is to return to the glory ancient Chinese enjoyed, in which China could paint other countries as barbarians with no culture, while depicting themselves as the "Central Country" with a highly developed culture.

In explaining a similar position, taken by the New Party, which represents a small portion of Taiwanese who endorse the PRC's vision of a "Greater China," Garver (1997) remarks that the New Party ". . . longs for the day when China is accorded its rightful place as one of the leading nations of the world. It desires a powerful Chinese nation-state and derives a deep satisfaction from identification with that would-be state" (pp. 19–20). This "would-be state" could conceivably help resolve feelings of frustration, impotence, and humiliation that Chinese have experienced (Tu, 1994b).

Hence, the construction of Chinese identity, which lies at the core of the call for unification, demands interpretation at local, national, and international levels, leading to ". . . negotiations about identity and manipulations of policy by the governments of China and other nations" (D. Wu, 1994, p. 149). In a broader sense, the call for unification involves not only people in Taiwan and China, but may also include all people of Chinese descent, even those living overseas.[2]

It is interesting, however, to note that while the PRC struggles to maintain its unity and centrality in light of its turbulent modern history, it has not been entirely successful in its social and economic development. On the contrary, other, more peripheral, Chinese communities, such as Taiwan and Hong Kong, have claimed increasingly greater status in the world economy. Tu (1994b) notes that such a situation changes the focus from a would-be central political entity to peripheral entities, and thus permits these peripheral elements to undermine the center's political effectiveness by setting the incipient alliance's economic and cultural agenda. This geopolitical shift may in fact, Tu concludes, ". . . be a blessing in disguise for the emergence of a truly functioning Chinese civilization-state" (p. 16).

The Economic and Cultural Connectedness

From a more pragmatic perspective, regardless of how people in Taiwan and China resolve their political disagreements, they have become more and more economically dependent upon each other (Chang, 1995; Garver, 1997; Mowlana, 1996). Capital investment, transfer of technological and managerial expertise, relocation of factories, and so on, have taken place between Taiwan and the coast of China on a large scale and with unprecedented speed. For example, Taiwan's investment in China was between $2 billion and $5 billion in 1991, between $7 billion and $9 billion in 1993, and between $14 billion and $20 billion in 1994 (Chang, 1995; Garver, 1997).

In addition to economic exchanges, cultural exchanges among Chinese people in Hong Kong, Taiwan, and China have also occurred at various levels and at an accelerated pace. An estimated one million residents of Taiwan visit China each year, bringing about two billion dollars annually to their families and relatives on the Mainland. Moreover, approximately fifteen million pieces of mail, phone calls, and telegrams, are exchanged each year (Chang, 1995). Equally powerful is the spread of popular culture from Taiwan and Hong Kong to China (Gold, 1993; Harding, 1993). Hence, despite the political puzzles and the lack of any legalized or formal rules of interaction between China and Taiwan, economic and cultural exchanges are made possible based upon the traditional Chinese interpersonal ties and informal connections (Yahuda, 1993).

With Taiwanese making ever greater investments on the Mainland for longer periods of time, the PRC attempts to use Taiwan's economic dependence on China to prevent it from becoming an independent nation, perhaps in the hope that Taiwanese businesspeople will pressure the ROC government to make concessions (D. Wu, 1994). According to Regnier, Niu, and Zhang (1994),

> It does not mean that central or local governments do not play an important role as facilitators, but the major actors are businessmen, technicians, scholars, and scientists, and various types of nongovernmental organizations. In the end of any removal of political barriers in the future, this subregion could become a single market, tentatively paving the way for Chinese unification. (p. 342)

Given their economic as well as cultural interdependence, Taiwan and China may reach some kind of political compromise in the future (Regnier et al., 1994), particularly since the image of the Chinese merchant has changed from that of a "trader" to that of a "financer" (Tu, 1994b). Regardless of how China and Taiwan succeed (or not) in resolving the thorny political problem of "one China," for some observers, a kind of economic and cultural "one China" is already taking shape. Many hope that these increased exchanges will naturally nurture and affirm the necessity for Chinese unity.

Other analysts, however, point to a disjunction between the economic and political orders, as China and Taiwan try to negotiate a "one China" policy (Harding, 1993; Yahuda, 1993). While China hopes that greater economic interdependence will hasten Taiwan's political integration with the Mainland, of equal importance is the Taiwanese idea that such interdependence could dilute China's eagerness to encompass Taiwan, simply because the PRC cannot afford to lose the economic benefits

Taiwan has brought to their association (Harding, 1994). Greater economic dependence would thus prevent China from further minimizing Taiwan's power on the international political stage.

THE CONCEPT OF THE "GREATER CHINA"

Another idea, close in meaning to the notion of "one China," is the idea of "Greater China." The term "Greater China" (dazhonghua in Chinese) appears to have become popular in the late 1980s,[3] as people sought to account for the increased economic interaction among various Chinese political entities; to identify the potential benefits such an integrated system might bring them; and to understand the return of Hong Kong to China (Huang, 1992; Lin & Robinson, 1994; Shambaugh, 1993).

Just which political entities might comprise "Greater China" is not precisely clear. "Greater China" might include Taiwan, China, Hong Kong, and Macao; on the other hand, the elements of a "Greater China" system could shrink to include only Taiwan and the coastal provinces of China, or expand to include any country which has a large Chinese presence (such as Singapore or Malaysia), or, most broadly conceived, could include not just "Chinese" countries, but any overseas Chinese (whether in America, Europe, or any other part of the world) (Harding, 1993; Robinson & Lin, 1994).[4] As economic ties among the potential members of "Greater China" vary from one period to another, so do the boundaries of the proposed community (Harding, 1993). For example, Chang (1995) identified three progressively expanding forms of "Greater China": Greater China I includes Taiwan, Hong Kong, and Guangdong and Fujian, two coastal provinces of China. Greater China II includes the entities in Greater China I, as well as all other provinces in China. Greater China III includes all entities in Greater China II, as well as all overseas Chinese.

Most commonly, however, the concept of "Greater China" is used to refer to economic exchanges among Taiwan, Hong Kong, and the southern provinces of China (primarily Guangdong and Fujian), which are said to form an economically integrated geographical area with approximately 120 million inhabitants (Chang, 1995). The increasing use of the term "Greater China" reflects an attempt to transcend national boundaries and to build some integrated system among people of Chinese descent. The emergence of this idea is timely, as Chang (1995) points out,

> The nation-state now strikes many as a "regressive force," incapable of serving progressive economic purpose. What modern conditions

require are larger functional entities and "global tribes" or "global civilizations" to serve them. (p. 967)

Moreover, the likelihood of forming a transnational Chinese community is now considered greater, given the widespread belief among Chinese that Chinese are better than non-Chinese in doing business with other Chinese, based upon language and cultural affinities, kinship, family, and ancestral ties, together with long-held cultural principles regarding interpersonal connection (Chang, 1995; Harding, 1993, 1994; Huang, 1992).

The concept also subsumes many related concepts, such as the "Chinese Common Market," "South China Economic Circle," "Chinese Civilization Community," and "Chinese Federation," each entity pointing to different aspects of political integration, cultural ties, or economic connections (Harding, 1993, 1994; Huang, 1992). These alternative formulas not only differ with respect to the scope of territories to be included, but also in terms of the degree to which there should be formal integration. Among these various formulations, however, "Greater China" is frequently the descriptor of choice, precisely because of its vagueness, ambiguity, flexibility, and all-encompassing nature. As Harding (1993) aptly summarizes,

> Not only does it have a simplicity and familiarity that its rivals lack, but its very vagueness is one of its greatest virtues. In contrast to more specific terms, the phrase "Greater China" does not imply any definite geographic boundaries, is not limited to any specific set of interactions, and does not refer to any definite structures or institutions. It accurately captures both the broad scope and the uncertain outcome of the phenomenon in question. (p. 683)

The metaphor "Greater China" has attracted a great deal of attention among scholars, journalists, business people, and politicians.[5] Though it is important to Chinese, the concept is used much more extensively by American academics than by the Chinese themselves, even though the more contemporary use of the term was initiated by Chinese scholars in the 1970s (Harding, 1993). While American academics use "Greater China" more in terms of its economic connotations, Chinese cannot detach themselves from associating this term with the eventual question of political unity and cultural and ethnic ties among various national entities which, for the moment, remain politically separate (Yahuda, 1993).

Whether one adopts the more "narrow" or "wide" definition of "Greater China," and whether one chooses to subscribe only to its economic aspects or to a broader sense of integration, it is still the case that much is revealed by the fact that there exist so many concepts to describe various levels of Chinese integration. Regardless of how one chooses to construe "Greater China," it remains true that some Chinese urgently want to present some kind of unified front guided by their sense of "Chinese-ness." Building an integrated, economic "Greater China" may thus be a prelude to a new, all-encompassing Chinese "nation" claiming as its citizens one out of every four people on earth.

For these reasons, the concept of "Greater China" is itself controversial, encompassing contrasting connotations. Not only is it impossible to define the parameters and identify boundaries of this concept, but it is extraordinarily difficult to specify its efficacy, utility, implications, and even appropriateness (Shambaugh, 1993).

> To some observers, it denotes the emergence of a dynamic new marketplace, offering greater economic opportunity to Chinese and foreigners alike. To others, in contrast, it evokes the image of an aggressive China on the march, seeking to extend its sovereignty over territories and people it does not presently control. Notwithstanding the debate over the appropriateness of the term, "Greater China" undeniably provides a succinct and vivid way of characterizing the growing interaction among the members of what Joel Kotkin has called the global Chinese tribe. (Harding, 1994, p. 235)

The notion of "Greater China" has pragmatic implications for both Taiwan and China. It is strategically used by the Taiwanese government to gain more international space, autonomy, and self-identity (Lampton, 1994). As the Taiwanese seek to appease China with respect to their willingness to unify, it is argued that China should not prevent Taiwanese from expanding their international political status. Taiwan has also been on guard against becoming China's subordinate, as well as avoiding political connotations that may damage the status of the ruling party, KMT. China, on the other hand, applies this concept carefully to avoid being seen as weakening in their resolve to unify Taiwan and the Mainland (Yahuda, 1993).

To put it simply, the sense of Chinese community, as articulated in the concept of "Greater China," is filled with contradiction, inconsistency, and multiple layers of meaning that prove complex, muddy, and even obscure. From an economic standpoint, "Greater China" might become a necessity, as more and more mutual interdependence develops between Taiwan and China. From a political perspective, the possibility of a "Chinese community" depends not only upon negotiations between China and Taiwan, but also upon other nations. In the view of some, the idea of "Greater China" can be used to point out the promise of a glorious future, even though this may in the view of others appear to be an aggressive attempt to extend China's sovereignty. From China's point of view, "Greater China" is related to the notion of "one China" and thus is part of an attempt to live up to the responsibility of "the Chinese people," whereas Taiwanese may see such a concept as a product of political compromise, used only for the purpose of deferring China's aggressive attempts to expand its sphere of influence.

Although the development of the concept of "Greater China" helps us understand the increased interactions among various Chinese communities, whether the concept will become a reality is still an open question, particularly given the divergent views between China and Taiwan. As Harding (1993) points out, many factors—for example, different sociopolitical systems, divergent living standards, protection of the interests of the countries involved, concern over reactions of other non-Chinese countries, and so on—may lead to the disintegration, rather than integration, of the "unified" entity, particularly since there exist neither formal diplomatic ties among these countries nor agreed-upon institutional rules to guide their economic exchanges (see also Harding, 1994; Yahuda, 1993).

While the many variables affecting this situation make it difficult, if not impossible, to predict the future direction of Chinese community among China, Taiwan, and other Chinese entities/groups (based upon interdependent economic activity generated by people of various Chinese societies) it is possible that a "Greater China" would become Asia's most powerful country and even a global superpower (Robinson & Lin, 1994). Whether there will be unity or disintegration, or even formal negotiation, many Chinese prefer to maintain the current situation and let history take its own course.

CHINESE COMMUNITY AND ITS IMPLICATIONS FOR A GLOBAL SOCIETY

Fundamentally, however, these varying perspectives all point to "the construction or revival of economic, political, and cultural ties among dispersed Chinese communities around the world as the political barriers to their interaction fall" (Harding, 1994, p. 235). While no one can predict with certainty that there will be an integrated "Chinese community," even the possibility of such a future state serves as a catalyst in the generation of alternative world orders.

First, the idea of "Greater China" allows Chinese to mediate and find a balance point between locality and globalness—to Chinese, the sense of community is local, since they share the same language and similar cultural traditions, but at the same time global, since it transcends the boundaries of various political entities. It reflects the effort of "reaching in" (getting in touch with "other" Chinese) as well as "reaching out" (going beyond one's national and ideological limitations), thus compelling Chinese to redefine their sense of identity with each other in mind (Yahuda, 1993). For Chinese of various political entities to build a larger Chinese community, they must simultaneously affirm their "Chinese-ness," while at the same time adopting a "global mindset" toward other Chinese, since these "other" Chinese may not share similar social and political backgrounds (Chen & Starosta, 1998).

The concept of "Greater China" apparently aids in such a vision, if not politically or economically, at least rhetorically. As Huang (1992) passionately advocates,

> After separate developments [of China and Taiwan] for forty or fifty years, Chinese, [are now] in a mutually stable and peaceful world facing the 21st century, and will have a "golden opportunity" to use methods such as "one country, two systems" or "Chinese community (Economic Circle)," to accomplish their historical responsibility of "China unification, and the re-creation of Hua Xia"! (p. 210)

While the notion of belonging to one centralized authority has always been a signature feature of Chinese political discourse (Pye, 1985), the call for the formation of a "Greater China" has been made more likely, since "The post-Cold War euphoria over possibilities for peace dividends from reduced militarism subsided quickly in the face of renewed struggle for new world orders" (Mowlana, 1996, p. 100). China no longer needs to play the strategic role of managing a balance between America and the Soviet Union, and thus should be able to concentrate on its own economic development and its relation within the sphere of Pacific nations. Similarly, Taiwan should find more opportunities to develop its international status, as more countries are willing to establish some form of relations (Yahuda, 1993). In the face of a new world order which remains to be established, the potential for a Chinese community must start with Chinese people establishing a new sense of "Chinese-ness"—that is, both racially and culturally, but not politically—by building interlinkages among themselves and with other, non-Chinese, countries.

Second, although the notion of having a "Greater China" may be seen as an attempt to fulfill the dream of some Chinese for only one Chinese nation—or at least, having somewhat closer relations among various Chinese political entities—the vision of such unity, characterized by the interconnections built by people of China and Taiwan, will affect how they are perceived by other nations, particularly Japan, their Southeast Asian neighbors, and America. Coincident with the end of the Cold War, the emergence of "Greater China" has led the outside world to react to the potential configuration of Chinese countries in a specific context (Yahuda, 1993).

Lampton (1994) argues that the concept of "Greater China" is used less to predict the direction of the future, than to provide help with current situations. In the case of the United States, for example, use of the term implies that "one cannot adopt policies (e.g., tariff policies, technology transfer policies, or sanctions) aimed at one portion of 'Greater China' without considering the effects on the other parts . . ." (p. 266). In other words, other nations, having to abide by the prevailing "one China" policy, must take into account the balance of the system composed by China and Taiwan when conducting economic transactions with either of the two political entities. A reconfiguration is thus unavoidable.

Third, the idea of "Greater China" allows Chinese to redefine the concept of a global society from their own perspective, rather than having to adopt perceptual frameworks from the West. This is particularly important, since, as Tu (1994b) points out, the culture of China has been playing a marginal and negligible role in the discourse of the "global village." Constructing a more interdependent Chinese community—whether on the basis of economy, culture, or politics—helps motivate Chinese to claim a greater stake in the shaping and definition of a global society. This attempt is appropriate, given the turn of global political situations:

> The first year of the 1990s clarifies the shape or the future of international relations and how they would influence Asia and Greater China. The 1991 Gulf War, the final demise of communism in the Soviet Union and the subsequent disintegration of that country, and the breakdown of domestic societies on several continents are one factor in that future. They signified the final breakdown of the cold war bipolar system centered on American-Soviet power and ideological rivalry throughout the globe. (Robinson & Lin, 1994, p. 456)

It seems plausible to view the current vacuum in international politics as an opportunity for Chinese communities to step to center stage in world affairs. In the past, the West, particularly America, has had various means of power to influence the shaping of global society. The

Cold War of opposition between America and the Soviet Union has to a large extent set the stage for people to contemplate the meaning of global society. At the present time, with one-fourth of the world's population Chinese, the Chinese slogan, "the 21st century will be the Chinese century," reflects the Chinese desire to play a central role in defining global society, thereby enabling them to claim a greater role in world affairs.

The possibility of creating a Chinese community demands the rebalancing of power struggles within the global society, particularly in view of the changes wrought by the end of the Cold War. Regionalism and universalism will continue to contest each other, as we see the formation of alliances among countries in North America, in Europe, among nations espousing the Muslim faith, and finally, in the possibility of a "Greater China." As Shambaugh (1993) writes, "As the new millennium approaches, one has the sense that the world is in transition from one epoch to another. Among the new realities of our era is the emergence of 'Greater China'" (p. 653). The concept of "Greater China" also

> . . . joins other phrases—"the new world order," "the end of history," "the Pacific Century" and the "clash of civilizations"—as part of the trendiest vocabulary used in discussions of contemporary global affairs. (Harding, 1993, p. 660)

As can be observed from the case of the Chinese, the intersection between sense of community and global society is never simply a question of volition or a matter of ideals, but an ongoing enterprise that can never be separated from pragmatic, political struggles. As Young (1996) puts it,

> We have witnessed the end of the cold war, the growth of global rather than merely local problems of pollution, increased trade and economic interdependence, greater levels of travel and migration, accelerating population growth, and increased electronic means for communication across planetary distances. (pp. 1–2)

The new world order still remains to be established. Of course, we still have not dealt with the questions of whether the formation of a Chinese community is possible, what would it look like, and what means would allow it to take shape. If the boundaries between nation-states were to be dissembled, what would allow people to move beyond the confines of physical locality to manage new alliances? Will the new Chinese community be characterized by extensive economic exchanges, built upon people's faith in their common Chinese cultural heritage, and aided by modern communication technology that make it possible for physically distant Chinese to share a sense of proximate involvement with each other (Chen & Starosta, 1998)? Or will it merely be an empty slogan, a call made possible by global media, and become reality only in the context of brute realpolitik? Here we find politics and economics intimately intertwined with the technology and global media that characterize the new millennium. As we move into the twenty-first century, the discourse of Chinese community will continue to impinge upon the formation and meaning of our "global society"—we are only at a beginning point in our contemplation of these possibilities.

NOTES

1. Other alternative formulas have been proposed throughout history, including "one country, one system," "one country, two governments," and "one country, many governments" (Harding, 1993).

2. Tefft (1994) notes that, apart from apparent economic gains, the vision of a new China proposed by Deng Xiaoping's "open door" policy has aroused much emotion associated with a reinvigorated China, and has lured many overseas Chinese to invest in the Mainland (see also Chang, 1995).

3. Harding (1993) notes that references to "Greater China" appeared as early as 1930. More contemporary usage of the term in Chinese language appeared in the late 1970s, and in English around the mid- to late-1980s.

4. Scholars and others debate about what political entities/communities/countries should be included in the "Greater China" concept (Harding, 1993). For example, Huang (1992) notes that "Greater China" should include only China, Taiwan, and Hong Kong. Since Singapore is not composed entirely of ethnic Chinese, it should not be included in "Greater China." Others contend that "Greater China" should be expanded to include not only Singapore (given that its principal ethnic group is Chinese), but also Chinese in Southeast Asia, and even overseas Chinese in Europe or America. Aside from this more economically oriented conception of the "Greater China," Tu (1994b), in explaining his notion of a "cultural China," includes non-Chinese who are interested in an intellectual understanding of Chinese

(see also Harding, 1993). It is also interesting that, as Yahuda (1993) points out, "Paradoxically it seems easier for outsiders to delimit the concept geographically than it is for the Chinese themselves. More is involved than sheer practicalities in deciding whether to exclude Singapore or different parts of the mainland" (p. 690).

5. *China Quarterly*, for example, published a special issue in 1993 to address issues relating to "Greater China," based upon a previous conference held in Hong Kong.

REFERENCES

Chang. M. H. (1995). Greater China and the Chinese "global tribe." *Asian Survey, 35*(10), 955–967.

Chao, L., & Myers, R. H. (1998). *The first Chinese democracy: Political life in the Republic of China on Taiwan.* Baltimore, MD: The Johns Hopkins University Press.

Chen, G. M., & Starosta, W. J. (1998). *Communication and global society: An outline of research.* Paper presented at the annual meeting of the National Communication Association, New York.

Chen, L.C., & Reisman, W. M. (1972). Who owns Taiwan?: A search for international title. *The Yale Law Journal, 81*(4), 599–671.

Clark, R. P. (1997). *The global imperative: An interpretive history of the spread of humankind.* Boulder, CO: Westview Press.

Copper, J. F. (1996). *Taiwan: Nation-state or province*? Boulder, CO: Wetview.

Garver, J. W. (1997). *Face off: China, the United States, and Taiwan's democratization. Seattle*, WA: University of Washington Press.

Gold, T. B. (1993). Go with your feelings: Hong Kong and Taiwan popular culture in Greater China. *China Quarterly, 136*, 907–926.

Harding, H. (1993). The concept of "Greater China": Themes, variations, and reservations. *China Quarterly, 136*, 660–686.

——. (1994). Taiwan and greater China. In R. G. Sutter & W. R. Johnson (Eds.), *Taiwan in world affairs* (pp. 235–275). Boulder, CO: Westview Press.

Huang, Z. L. (1992). *China moves toward the twenty first century.* Hong Kong: Joint Publishing (H.K.), Co., Ltd.

Kotkin, J. (1993). *Tribes: How race, religion, and identity determine success in the new global economy.* New York: Random House.

Lampton, D. M. (1994). Commentary. In R. G. Sutter & W. R. Johnson (Eds.), *Taiwan in world affairs* (pp. 265–270). Boulder, CO: Westview Press.

Lin, Z., & Robinson, T. W. (1994). Preface. In Z. Lin & T. W. Robinson (Eds.), *The Chinese and their future* (pp. xiii-xv). Washington, DC: The AEI Press.

Morgan, P. M. (1993). Taiwan's security interests and needs. In G. T. Yu (Ed.), *China in transition: Economic, political, and social developments* (pp. 3–22). Landam, MD: University Press of America.

Mowlana, H. (1996). *Global communication in transition: The end of diversity?* Thousand Oaks, CA: Sage Publications.

Naisbitt, J. (1994). Global *paradox*. New York: Avon.

Pye, L. W. (1985). *Asian power and politics: The cultural dimensions of authority*. Cambridge: Harvard University Press.

Regnier, P., Niu, Y., & Zhang, R. (1994). Toward a regional "block" in East Asia; Implications for Europe. In B.-j. Lin & J. T. Myers (Eds.), *Contemporary China and the changing international community* (pp. 331–347). Columbia, SC: University of South Carolina Press.

Robinson, T. W., & Lin, Z. (1994). The Chinese and their future. In Z. Lin & T. W. Robinson (Eds.), *The Chinese and their future: Beijing, Taipei, and Hong Kong* (pp. 445–472). Washington, DC: The AEI Press.

Shambaugh, D. (1993). Introduction: The emergence of "Greater China" *China Quarterly, 136*, 653–659.

Tefft, S. (1994). The rootless Chinese: Repatriates transform economy, yet endure persistent resentment. *Christian Science Monitor*, March 30, p. 11.

Tu, W. M. (1994a). Preface to the Stanford edition. In W. M. Tu (Ed.), *The living tree: The changing meaning of being Chinese today* (pp. v-x). Stanford, CA: Stanford University Press.

——. (1994b). Cultural China: The periphery as the center. In W. M. Tu (Ed.), *The living tree: The changing meaning of being Chinese today* (pp. 1–34). Stanford, CA: Stanford University Press.

Wang, Y. S. (1985). The Mainland-Taiwan issue and China's reunification: Road toward reconciliation. In Y. S. Wang (Ed.), *The China question: Essays on current relations between mainland China and Taiwan* (pp. 35–49). New York: Praeger.

Wei, Y. (1994). The effects of democratization, unification, and elite conflict on Taiwan's future. In Z. Lin & T. W. Robinson (Eds.), *The*

Chinese and their future: Beijing, Taipei, and Hong Kong (pp. 213–240). Washington, D.C.: The AEI Press.

Weng, B. S. J. (1995). The evolution of a divided China. In Z. Lin & T. W. Robinson (Eds.), *The Chinese and their future: Beijing, Taipei, and Hong Kong* (pp. 345–385). Washington, D.C.: The AEI Press.

Wu, A. C. (1994). Taipei-Peking relations: The sovereignty issue. In B. J. Lin, & J. T. Myers (Eds.), *Contemporary China and the changing international community* (pp. 187–198). Columbia, SC: University of South Carolina Press.

Wu. D. Y. H. (1994). The construction of Chinese and non-Chinese identities. In W. M. Tu (Ed.), *The living tree: The changing meaning of being Chinese today* (pp. 148–167). Stanford, CA: Stanford University Press.

Yahuda, M. (1993). The foreign relations of Greater China. *China Quarterly, 136*, 687–710.

Young, R. (1996). *Intercultural communication: Pragmatics, genealogy, deconstruction.* Clevedon, England: Multilingual Matters Ltd.

CHAPTER FIVE
The Challenge of Diversity in Global Organizations

Jensen Chung
San Francisco State University

A dialectical relation between the global and the local often emerges in the process of globalization. In a globalizing society, the trend of globalization often enhances the needs of localization. As Naisbitt (1994) points out, although people want to come together to trade much more freely, they want to be independent politically and culturally. In other words, the more people are bound together economically, the more they want to assert their own distinctiveness, independence, and sovereignty.

A similar quandary exists in a global organization. A global organization "not only must do business internationally but also must have a corporate culture and value system that allow it to move its resources anywhere in the world to achieve the greatest competitive advantage" (Rhinesmith, 1996, p. 5). Rhinesmith considers a global organization to be concentrating primarily on achieving economies of scale through sharing resources. Because of the necessity of sharing resources, building a universal corporate culture becomes impelling to global organizations. In fact, it often is argued that part of the process of globalization is the integration of the organizational cultures. Rhinesmith (1996), for example, suggests forming integrated values, mechanisms, and processes to develop a global corporate culture. In these processes, leaders of the organization create throughout the company a synergy and shared culture. In addition, computer networking technology may also help centralize market forces and create a homogeneous economy culture (Harasim, 1993). Interdependence and frequent interactions among local or national organizations are key features, or even impetuses, of globalization. A universal corporate culture would reduce communication barriers in a diversified organization.

Along with the integration of an organization, especially a larger organization, however, comes a contradictory force: the need for segmentation. The segmentation of a global organization is also inevitable, because telecommunication and computer-mediated communication such as electronic mail have made information increasingly available. Individuals and local organizations or branches are thus empowered in turn promoting independence politically and culturally. Appadurai (1990) posits that the central problem of globalization is the dialectic tension between cultural homogenization and heterogenization. The contradiction between unification and independence may account for why multinational companies do not and cannot submerge the individuality of different cultures (Laurent, 1983). However, how members of global organizations face this contradiction remains to be studied. To that end, how organizational members from dominant and non-dominant cultures clash on a particular value needs to be investigated.

This chapter focuses on how co-cultures in a global corporation view the issue of assertiveness. This issue is brought to a focus mainly because of three reasons: First, assertiveness is potentially a point susceptible to clashes between major cultural values. As Rakos (1991) points out, the concept of assertiveness is "a manifestation of the *dominant* American value system that guides the behaviors of *whites*" (p. 13). He also notes that in the US, co-cultural members that report being less assertive than whites tend to be those with strong cultural identities, such as Mexican-Americans (Hall & Beil-Warner, 1978), Japanese-Americans and other Asian-Americans (Fukuyama & Greenfield, 1983), and Chinese-Americans (Sue, Ino, & Sue, 1983). Second, when assertiveness is a value cherished in a culture, it would appear to be a pervasive organizational issue involving day-to-day problems of interpersonal conflict, negotiation, problem-solving, leadership, and superior-subordinate communication. Assertiveness may then be a major criterion in assessing communication competence by lay persons in dealing with those problems. Third, in global organizations, members from numerous different cultures communicate. Assertiveness can mean confidence in one's own cultural heritages and standing up for these heritages. Assertiveness, therefore, is an element shaping cultural identities in global organizations, and maintenance of cultural identities is a challenge to the issue of diversity in the process of globalization.

ASSERTIVENESS AND CULTURE

Although the concept of assertiveness is culture bound, the conceptualization of the concept tends to be dominated by American cultural values. For example, Furnham (1979) observes that assertiveness is particularly a North American concept. Rakos (1991) considers assertiveness training in part "a product of America's enduring allegiance to individual freedom" (p. 13). He argues that throughout its history, the US has been a nation concerned with protecting individual rights, often at the expense of communal interests. Samovar and Porter (1995) view the meaning of assertiveness as dictated by the general American culture that exposes its members to countless hours of television talk shows glorifying verbal combat and that has more lawyers per 10,000 people (19.6) than any other culture in the world.

Other cultures have values that hold contrary views of assertive communication. For example, the emphasis of interpersonal harmony and confrontation avoidance in Confucianism-influenced cultures like those in China and Taiwan usually devaluates assertive communication. In Japan, the virtue of *enryo*, or self-restraint, is exercised to avoid causing displeasure (Lebra, 1976) and is more valued than asserting oneself. To the Japanese, assertiveness is considered as immature or selfish (Fukumura, 1995).

Latino Americans have a web of traditional values that promote humility, pessimism, and self-denial over the kind of individual ambition required for success (Sosa, 1998). The Spanish teach subservience in the name of good manners. Subservience is built into the polite Spanish idioms of response, such as the fatalistic *como Dios quiera* (i.e., however God wants it) and *mandeme* (i.e., command me) to express that they are ready to listen. O'Hara-Devereaux and Johansen (1994) note that in a hierarchical culture such as Mexico, a lower-status worker would normally be "out of order" if he or she were to question a senior or appear to pass information up the hierarchy. Rakos (1991) reports that respect for authority figures, parents or elders in particular, is a strong value in African-American (Mitchell-Jackson, 1982), Asian and Asian-American (Fukuyama & Greenfield, 1983; Hong & Cooker, 1984; Su, Ino, & Sue, 1983), Puerto Rican (Comas-Diaz & Ducan, 1985), Mexican-American (Grodner, 1977; Mitchell-Jackson, 1982), and Native American (Mitchell-Jackson, 1982) cultures. The value of respect is so strong in these cultures that even empathic assertions to such persons are likely to be judged to be highly inappropriate (Rakos, 1991).

As a consequence of the differences in cultural values associated with assertiveness, Americans are reported to be significantly more assertive than Koreans, Finns, and the Japanese (Park & Kim, 1992). Studies indicate that Americans are more active, assertive, verbal, and direct than Japanese, who are more passive, restrained, reserved, modest, nonverbal, and indirect (Barnlund, 1975; Barnlund & Araki, 1985; Nomura & Barnlund, 1983; Thompson, Kloph, & Ishii, 1991; Tomioka, 1994). Even in the US, Asian Americans are reported to be less assertive (Zane, Sue, Hu, & Kwon, 1991).

Rich & Schroeder (1976) propose that assertive behavior is "the skill to seek, maintain, or enhance reinforcement in an interpersonal situation through the expression of feelings or wants when such expression risks loss of reinforcement or even punishment" (p. 1082). The majority of theory and research on assertion has been focused on negative situations, such as giving criticism, disagreeing, and refusing requests (Wilson & Gallois, 1993). Because the conceptualization of assertiveness is very much American-oriented, assertiveness is treated both in academia and in training practice as being more psychological than cultural. For example, Alberti and Emmons (1990) argue that non-assertive behaviors generally are anxiety-based. Their studies consider non-assertiveness as a matter of control. In other words, we may fail to control our relationships or ourselves when responding to certain situations, personalities, topics, and behaviors. The authors state that some possibilities might make one hold back from asserting oneself: laziness, apathy, feelings of inadequacy, fear of being considered unworthy, unacceptable, fear of hurting another person or making him or her angry, fear of getting no reinforcement, not knowing how to accomplish a desired goal, or feeling that if you don't do it, someone else will. None of these possibilities are cultural.

This chapter focuses on identifying cultural reasons of the perceived assertiveness problems of Taiwanese-American subordinates working in global corporations and discusses its implication to the dialectical relation between diversity and identity in the global organization. The author designed an interviewing protocol by using the framework based on research in the fields of interpersonal communication, intercultural communication, and leadership communication. The protocol includes five open-ended questions regarding assertiveness in a global organizational setting (e.g., What do you think are the reasons why you are perceived as assertive or non-assertive?). Where there were confusions about the definition of assertiveness, the following definition was provided as a reference: Assertiveness refers to how well you act or speak up to show who you are, what you want, and what you think, especially when you need to protect your rights.

The author interviewed 29 high-tech engineers and held three focus-group discussions. Seven of the 29 participants went to both the interviews and the focus group meetings. All of the participants were born and raised in Taiwan, which is considered strongly influenced by Confucianism. They had lived in the United States for an average of 11.5 years. Among them, 21 were male; 8 were female. Their ages ranged from 25 to 55 with an average age of 37. All of them had subordinate status, although one fifth of them were of managerial rank. Seventy-one percent of the participants' superiors at work are Americans. The engineers were all U.S. citizens and were working in the high-tech industry in the Silicon Valley area of California. They belong to seven global corporations, which, with the exception of two, are all American-based. Pseudonyms of the participants are used in this chapter.

CULTURAL INFLUENCES ON PERCEIVED NON-ASSERTIVENESS

The results show that ten themes emerged from transcripts of interviews and focus group meetings. These themes explain why they were perceived as non-assertive. Most themes are illustrated with narratives of some of the 29 participants. The 10 themes are social role acknowledgment, reciprocity expectation, expected relationship duration, face consciousness, confrontation style, verbal frequency, social context, linguistic fluency, social rule familiarity, and modesty. They are discussed in the following section.

Social Role Acknowledgment

In Taiwan, like other Confucianism-influenced cultures, people show respect with verbal and nonverbal messages to those who play certain social roles. These social roles include those who are considered elders, having higher social status (e.g., the gentry class), being in higher social hierarchy (e.g., relatives of earlier generations, superiors at the workplace), or having certain socially respected professions (e.g., public functionaries or teachers). In a culture where such respect is not explicitly shown with verbal or nonverbal messages, the practice of acknowledging a communication partner's social roles can be construed as a lack of self-confidence or non-assertiveness.

Hofstede's (1980) theory of power distance can partly describe the difference in the degree of acknowledging social roles. According to Hofstede, power in low-power-distance cultures is relatively more equally distributed, and status is less marked. People in this kind of culture tend to challenge inequality in society, and subordinates in an organization consider job superiors to be "just like myself." U.S., Canada, New Zealand, Austria, and Israel are in this category. In contrast, people in high-power-distance cultures, such as those in Singapore, Taiwan, and the Philippines, tend to accept the unequal power. Hofstede's theory, however, cannot fully explain the practice of role acknowledgment, because in Confucianism-influenced cultures power is not the only criterion assessing whether one should show respect or deference to one's communication partner. It is the social role that serves as the main criterion for individuals' behaviors. This is reflected in a major theme emerging from the results of this study: showing respect for superiors is a polite behavior according to the teachings in their (Taiwanese) homeland, but in this (American) culture, the politeness is often perceived as a lack of confidence. The situation is especially true when the nonverbal messages are communicated differently. For example, the following cues showing good manners in many East Asian cultures can be interpreted in the West as appearing timid to the superior: bowing down too much, nodding too often, smiling too constantly, and shifting eye contact away from the boss too often.

Reciprocity Expectation

Confucianism-influenced employees expect superiors and subordinates to help each other and reciprocate each other's favors. Superiors do not take for granted obedience and respect from subordinates, and subordinates expect superiors to reciprocate by supporting, advising, and protecting them. Most respondents in this study indicate that, with the expectation of future reciprocation, being taken advantage of by superiors is considered a triviality and is not worth the hassle of a fight. Behaviors displayed by this attitude may be interpreted by members of other cultures as fear of defending their own rights or of speaking their own feelings.

According to Hofstede (1980), in individualistic societies (e.g., U.S., Britain, and Canada), personal goals take precedence over group goals, while greater emphasis is placed on needs and goals of the in-group rather than oneself in collectivistic societies. Sacrificing one's right to accommodate the superior or peers is a norm in collectivistic societies but may be interpreted as lacking assertion in individualistic societies.

Expected Relationship Duration

The way these Taiwanese-American engineers develop their interpersonal relationships may influence the perception of assertiveness. As Chen & Chung (1994) point out, it takes longer to develop an interpersonal relationship in East Asian cultures and when there are conflicts, those in the relationship rely more on the established relationship—sometimes called *guan shi* in the Chinese language, to help resolve the conflict. Before maturing the relationship, however, it is more difficult for them to engage in joint transactions or collaborations, but they can tolerate conflict more easily with the hope of future understanding and thus closer relationship. The tolerance is often interpreted as a non-assertive "grin and bear it." The comments of Mr. Chu indicate this tendency:

> During the two months when I was an expatriate in an Asian country, I joined a discussion group of five engineers in three countries, including my project leader in the States. We exchanged ideas through e-mail and teleconferences to solve a programming problem. When I returned to the States, I found that my project leader claimed in his report that a new idea for allowing our new computer to enter the hibernation mode was his. He didn't give me any credit at all, even though everyone in the discussion group knew that it was an idea I initiated. Colleagues urged me to protest. I thought that we would be colleagues for a long time, so I shouldn't bother to fight because of a small advantage he took. A month later, he took an offer from another company and left us. Through my friends in his new company, I learned that he got the lucrative offer because of the idea I initiated during the discussions.

Blake & Mouton (1978) posit that when handling conflicts, people consider two options: cooperativeness or assertiveness. Cooperativeness is the concern for the relationship with the conflicting party, and assertiveness is the means for achieving one's personal goals or interests. When our concern for the relationship is high, we back away from conflict. This withdrawing style in dealing with conflicts naturally makes one appear non-assertive. Chen and Chung (1994) further point out that East Asian cultures not only emphasize developing a relationship before carrying out tasks, but also encourage developing long-term relationships. Such emphases naturally encourage people to take a non-assertive position in dealing with conflict, as shown in Mr. Chu's case.

Face Consciousness

"Face" is a human relationship concept in East Asian cultures that makes people less aggressive in asserting their right or hurting others' esteem. Face is to preserve self-esteem or others' dignity. To avoid causing other people to "lose face," one must use indirect or ambiguous language to refuse or to accuse. As Yum (1988) observes, "indirect communication helps to prevent the embarrassment of rejection by the other person or disagreement among partners, leaving the relationship and the face of each party intact" (p. 374).

Several participants in the interviews and group discussions attested this point. They said they were often criticized at meetings. Although they were confident they were not mistaken, they did not rebuke for fear of embarrassing the critics. They hoped that the critics would later get to the point, so they could then provide the critics with facts. They said that in engineering, especially in the computer industry, numbers speak louder. Needless to say, this attitude makes a case for assertiveness. When harmony is a goal in interpersonal relationships, endurance, suffering, and tolerance are parts of the process in working toward that goal, but they also may be indications of non-assertiveness.

Confrontational Style

Another reason East Asians are perceived as non-assertive is that they take a non-confrontational style in dealing with conflict or a potential conflict. This style may result not only from the value of "face" and the orientation of long-term relationships, but also from the value of harmony.

East Asian cultures have been strongly influenced by Confucianism, Buddhism, and (especially in China and Taiwan) Taoism. Confucian principles set harmony as the goal of many teachings in interpersonal relationships; Taoism stresses the value of non-action and non-strength; and Buddhism maintains that everything in the earthly world eventually will pop like a bubble and, therefore, nothing is worth the hassle of a fight. From the Buddhist perspective, as Fisher and Luyster (1991) state, one must destroy one's ego to alleviate suffering and use communication in the service of truth and harmony.

The value of harmony has geared conflicting partners more toward the style of accommodation, instead of assertion or aggression. This value leads to the non-confrontational style in interpersonal conflict situations. A common style of dealing with conflicts that the participants suggested is asking an intermediary to convey their sentiment or complaints. When an intermediary is not available, they often withdraw. Even when being aggressively attacked in defining territory or allocating space, they do not confront the other party or counter-attack. Newer immigrants, in particular, believe that they gain much more space by withdrawing than by fighting for territory. A Chinese proverb most commonly justifies this argument and comforts them: "A step back, the sea is broad and the sky clear."

Having been in a non-confrontational conflict style, a person may regard criticism of someone's idea as a persona attack and hesitate to criticize a colleague's or superior's ideas, which may reinforce the perception of non-assertiveness.

Verbal Frequency

Hall (1976) categorizes cultures as being high-context or low-context. High-context cultures tend to maintain tradition and change little. Therefore, when people communicate, they can rely on information provided by the setting or the communicators, rather than verbal messages as in the low-context culture. Silence is often a sufficient message.

It is well known or stereotyped that East Asians or Asian Americans are reticent, reserved, passive, and non-assertive (Meredith & Meredith, 1966; Mayovich, 1972; Sue & Kitano, 1973). A Confucian saying, "Those who are clever in talks and flattery in facial expressions lack *Jen* (kindness)," has a clear imprint on the communication behaviors of East Asians and even East Asian Americans. In Taiwan, teachers have traditionally done all the talking in classes. In families, parents often told their children: "Children have ears, not mouths." As a result, Taiwanese, like most East Asians, are characteristically more reticent.

Assertiveness, indicated in the West by dominance, contentiousness, low anxiety, and refusals to be intimated by others, was reported to be correlated highly with verbal intensity, talkativeness, and good communicator style (Norton & Warnick, 1976). Several participants in this study, however, argued that simply because they are quiet does not necessarily mean that they cannot stand up for their own rights or that they cannot communicate.

Social Context

The older generations of Chinese believed that people should die, or at least, be buried in their hometown, otherwise, their spirit might wander around. "Fallen leaves return to the root" is a common Chinese saying used to denote that everyone eventually will need to return to their homeland. Although this "guest mentality" is diminishing in the process of immigration, most Taiwanese immigrants being marginalized in U.S. society still feel nostalgic about their homeland. This guest mentality is a part of their consciousness that influences their communication style, as indicated by Mr. Ho:

> When I was an elementary school kid in Taiwan, I visited my uncle's home once a year. My aunt had southern Taiwanese hospitality. She would use chopsticks to put food in my bowl every once in a while, even when I was full to the brim. Once, I was fed so much that I almost threw up. In my early months with the American company here in this country, I thought I had a similar experience: I was given so many new projects, and I took them all as training in the manager's good intention until I almost collapsed. As a Chinese proverb goes, "At home, rely on parents, away from home, count on friends." I used to treat colleagues as friends. Now, I know they are not; they often are competitors. Now, I also know my limit, and I let my manager know my limit. I speak up when given an unreasonable workload.

This guest mentality often exists in the early stage of immigrants' careers. Most participants in this study said they could manage to change this mentality, but were not sure if non-immigrant colleagues would notice this change.

While the above-listed social role acknowledgment, confrontational style, reciprocity, expected relationship development duration, and social context account for the reasons of non-assertiveness embedded in a deep cultural level, language barriers and social rules familiarity in the following sections appear at a relatively practical level and are more common to new immigrants.

Linguistic Fluency

Perceived failure to communicate assertively sometimes results from an inability to use vocabulary or expressions. When one cannot find a word to say, he or she may decide to be quiet. When inappropriate words are chosen for rejection or criticism, they may sound submissive or rude. Some participants described such dilemma: When critiquing a proposal, they might use expressions like "I disagree," "I don't think so," "that won't work," "it may fail, "that's not correct," and so on. They then found out that their native-speaker colleagues used words like "I have a concern that ...," "I wonder if ...," "Can you educate me about ..." and so on. But it might take them several years to discover these wording differences. Similar problems might occur when they reject requests from others. They wanted to avoid saying "no" to avoid appearing "rude," but they found themselves lacking word choice. On the other hand, they sometimes struggled to say something to defend their proposal, but their colleagues would rhetorically reject their argument, leaving them speechless.

Social Rules Familiarity

Social rules regulate and guide interpersonal relationships. They govern behavior in social contexts where assertive communication is recommended (Wilson & Gallois, 1993). Participants in this study point out some social rules that influence the perception of assertiveness. These rules include whether or when to start calling the superior by his or her first name, when to raise a hand to express opinions at meetings, when and how to interrupt, and what or which words one should use to critique more appropriately. All these questions demand an understanding of social rules in order to appear assertive. However, these rules are often difficult to learn or apply. The following narrative from another participant, who also has been in the U.S. for nine years, is an example:

> ... So, I decided to change my timid image. I also wanted to be assertive (laugh). I was going to start calling him (my boss) by his first name, but I was afraid that it would appear abrupt. Because his office is in another country, we normally communicate with E-mail. I started addressing him as Bob on E-mail last month. After that, he stopped addressing me in his e-mail—he used to address me by my first name in all his messages. I have been feeling so uncomfortable...It was hard for me to call my boss by his first name when I arrived, and now it's even harder to switch back.

The problem presented in this example is not unique to immigrants. Although rules for addressing vary from place to place and from organization to organization, it is difficult for immigrants to adjust to them because of cultural differences. Gallois and Callan (1991) report that there are a larger number of rules for unequal-status relationship than for equal-status ones. Inadequate understanding of social rules or poor habit of following social rules may cause employees raised in foreign cultures to appear non-assertive, especially to their superiors.

Modesty

Modesty is not a virtue unique to any particular culture. The emphasis on modesty in Confucianism, however, has a great influence on East Asian cultures. Being educated and working in the U.S. for various periods of time, almost all the Taiwanese participants still expressed their frequent internal conflicts between showing modesty and selling themselves. An example form Mr. Jun:

> Recently, I made a presentation to a group of Asian buyers. The responses were fantastic right after the presentation, and we sold our new design. However, when my supervisor congratulated me, I said, "Oh, well, I was just lucky. I didn't prepare the presentation long enough and, besides, I didn't have enough sleep last night." Reading the response in his face, I immediately regretted having said those modest words. My worry came true: I didn't get the "Bee of the Year Award" which, by rule, was presented to the hardest-working employee. Instead, it went to a colleague who just returned from half a year's leave of absence and made a presentation that was considered sloppy by my colleagues. Why? My supervisor nominated him instead of me.

All participants in the same group interview agreed upon Mr. Jun's belief about the relation between his modest remark and his failing to win the award. Admitting one's faults such as inadequate preparation apparently is equivalent to giving up one's right in an "assertiveness culture."

INTERPRETATION AND DISCUSSION

The emergence of the ten themes that explain perceived non-assertiveness of the Taiwanese engineers in this study can be attributed to three influences. First, the "self-awareness frame"—including role acknowledgment, reticence, and guest mentality. It dictates that with the cultural frame of reference in mind while communicating, participants self-monitor their role (e.g., as a junior or younger generation), position (e.g., as a subordinate), and status (e.g., as a guest). Second, the relationship values—including long-term relationship orientation and non-confrontational style. This influence shows that while communicating, participants expect to maintain an East-Asian style of relationship characterized by long-term orientation and harmony as the ultimate goal. Finally, communication skills—including language barriers and unfamiliarity with social rules. These influences on the perceived assertiveness are indicated in Figure 1.

Figure 1. Cultural Influences on Assertiveness of Minorities

	Influences on assertiveness	*Exemplary behaviors*
Self-awareness frames	Role acknowledgment Reticence Guest mentality	Show respect to superior Listen to the superior Polite to the host
Relationship values	Long-term relationship development Non-confrontation	Expect future understanding Slight gains from argument
Communication skills	Language barriers Unfamiliarity with social rules	Hard search for proper words Hesitate to change addressing

Chen & Starosta (1996) propose a model aiming at promoting interactants' abilities to acknowledge, respect, tolerate, and integrate cultural differences, so that they can qualify for enlightened global citizenship. The model represents a transformational process of symmetrical interdependence that can be explained from three perspectives: (a) affective, or intercultural sensitivity; (b) cognitive, or intercultural awareness; and (c) behavioral, or intercultural adroitness. Applying Chen & Starosta's model to examine the data in this study shows that "role acknowledgment" is a self-awareness process, and "familiarity with social rules," recognizing the differences in "confrontation style," "verbal frequency," "relationship development duration," and the concept of "reciprocity" are parts of a cultural awareness process. All these are connected with the cognitive (intercultural awareness) category in Chen & Starosta's model. Five items fall into the behavioral (intercultural adroitness) category: language barrier (message skill), physical context (behavior flexibility), social rule familiarity (interaction management), role acknowledgement (social skill), and face (empathy ability, social skill). Notice that some items of the influences can be classified in both cognitive and behavioral categories. Figure 2 shows the model of globalized organizational communication.

Figure 2. Model of globalized organizational communication

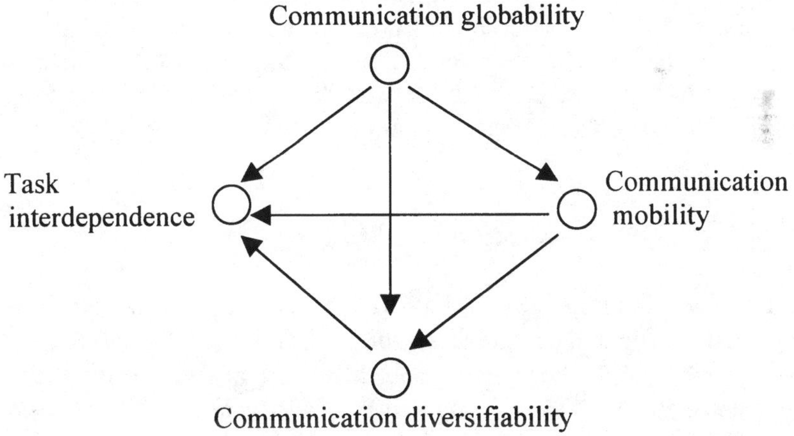

The ten influences on the assertiveness of the first-generation Taiwanese American high-tech engineers working in global corporations indicate that the perception of assertiveness is a clashing point issue. The conflict between globalization (or corporate value unification) and localization (or cultural diversification) does exist. In reference to the fact that the participants have lived in the U.S. for an average of 11.5 years, this study implies that it will take a long time to reduce or remove the influence of one's original cultural values. The greater geographical dispersion of global organizations makes the adaptation to different cultural values even more difficult. In addition, the increasing use of computer-mediated communication in global organizations may even complicate the issue of adaptation because of the lack of nonverbal cues and heavy reliance on text-based communication.

In conclusion, this study investigates how co-cultural members in global organizations view the perception of their "non-assertiveness problem." Their accounts depict the clashes of strikingly different values associated with assertiveness. Ten culture-related influences are identified. The results show that simultaneously paying attention to localization is a strong demand for modern organizations in the process of globalization. By the same token, global organizations need to acknowledge diversity in their efforts to unify corporate cultures, just as they need to expect segmentation during the process of integration from the perspective of organizational structure.

NOTE

Part of the research for this chapter was published in *Management Development Forum*, vol. 1, no, 2. An earlier version of this paper was presented at the 1998 annual conference of National Communication Association in New York. The author wishes to thank Debra Blackstone and Rudy Busby for their valuable comments.

REFERENCES

Alberti, R. & Emmons, M. (1990). *Your perfect right: A guide to assertive living*, 6th ed. San Luis Obispo, CA: Impact Publishers.

Appadurai, A. (1990). Disjuncture and difference in the global cultural economy. In M. Featherstone (ed.), *Global culture: Nationalism, globalization and modernity* (pp. 295–310). Newbury Park, CA: Sage.

Barnlund, D. (1975). Public self and private self in Japan and the United States. Tokyo, Japan: Simul Press.

Barnlund, D., & Araki, S. (1985). Intercultural encounters: The management of compliments by Japanese and Americans. *Journal of Cross-cultural Psychology, 16*, 9–26.

Blake, R. & J. Mouton (1978). *The new managerial grid.* Houston, TX: Gulf Publishing Co.

Chen, G. & Chung, J. (1994). Confucian impact on organizational communication. *Communication Quarterly, 42,* (2), 93–105.

Chen, G. & Starosta, W. (1996). Intercultural communication competence: A synthesis. *Communication Year book, 19.* Thousand Oaks, CA: SAGE.

Comas-Diaz, L. & Duncan, J. (1985). The cultural context: A factor in assertiveness training with mainland Puerto Rican women. *Psychology of Women Quarterly, 9,* 463–476.

Fisher, M. P. & Luyster, R. (1991). *Living religion.* Englewood Cliffs, NJ: Prentice-Hall.

Fukumura, Y. (1995) *Culture and gender in assertiveness and restraint among Japanese and Americans.* Unpublished Master's thesis, San Francisco State University.

Fukuyama, M. & Greenfield, T. (1983). Dimensions of assertiveness in an Asian-American student population. *Journal of Counseling Psychology, 30,* 429–432.

Furnham, A. (1979). Assertiveness in three cultures: Multidimensionality and cultural differences. *Journal of Clinical Psychology, 35,* 522–527.

Gallois, C. & Callan, V. (1991). Interethnic accommodation: The role of norms. In H. Giles, N. Coupland, & J. Coupland, (Eds.) *Contexts of accommodation: Developments in applied sociolinguistics* (pp. 245–269). Cambridge: Cambridge University Press.

Grodner, B. (1977). Assertiveness and anxiety: A cross-cultural and socioeconomic perspective. In R. Alberti (Ed.) *Assertiveness: Innovations, applications, issues.* San Luis Obispo, CA: Impact.

Hall, E. (1976). *Beyond Culture.* Garden City, NY: Doubleday. p. 79.

Hall, J. & Beil-Warner, D. (1978). Assertiveness of male Anglo and Mexican-American college students. *Journal of Social Psychology, 105,* 175–178.

Harasim, L. (1993). Networlds: Networks as social space. In L. Harasim (Ed.), *Global networks: Computer and international communication* (pp. 13–34). Cambridge, MA: MIT Press.

Hofstede, G. (1980). *Culture's consequences: International differences in Work-Related Values.* Beverly Hills: Sage.

Laurent, A. (1983). The cultural diversity of Western conceptions of management. *International Studies of Management and Organization, 13,* (1–2), 75–96.

Hong, K. & Cooker, P.G. (1984). Assertion training with Korean college students: Effects on self-expression and anxiety. *Personnel and Guidance Journal, 62,* 353–358.

Lebra, T.S. (1976). *Japanese patterns of behaviors.* Honolulu: The University Press of Hawaii.

Mayovich, M.K. (1972). Stereotypes and racial images—white, black and yellow. *International Journal of Social Psychiatry, 18,* 239–253.

Meredith, G.M., & Meredith, C.G.W. (1966). Acculturation and personality among Japanese-American college students in Hawaii. *Journal of Social Psychology, 68,* 175–182.

Mitchell-Jackson, A. (1982). Psychosocial aspects of the therapeutic process. In S. Turner & R. Jones (Eds) *Behavior modification in black populations: Psychosocial issues and empirical findings.* New York: Plenum.

Naisbitt, J. (1994) *Global paradox.* New York: William Morrow.

Nomura, N., & Barnlund, D.C. (1983). Patterns of interpersonal criticism in Japan and the United States. *International Journal of Intercultural Relations, 7,* 1–17.

Norton, R. & Warnick, B. (1976). Assertiveness as a communication construct. Human *Communication Research, 3,* 62–66.

O'Hara-Devereaux, M., & Johansen, R. (1994). *Globalwork: Bridging Distance, Culture, and Time.* San Francisco, CA: Josser-Bass Publishers.

Park, M. & Kim, M. (1992). Communication practices in Korea. *Communication Quarterly, 40,* (4), 398–404.

Rakos, R. (1991). *Assertive behavior: Theory, research, and training.* New York: Routledge.

Rhinesmith, S. (1996). *A manager's guide to globalization: Six skills for success in a changing world.* Chicago: Irwin Professional Publishing.

Rich, A. R., & Schroeder, H.W. (1976). Research issues in assertiveness training. *Psychological Bulletin, 83,* 1084–1096.

Samovar, L. & Porter, R. (1995). *Communication between cultures.* Belmont, CA: Wadesworth.

Sosa, L. (1998). *The American dream: How Latinos can achieve success in business and in life.* New York: Dutton.

Sue, D., Ino, S. & Sue, D. M. (1983). Non-assertiveness of Asian Americans: An inaccurate assumption? *Journal of Counseling Psychology, 30,* 581–588.

Sue, S., & Kitano, H. (1973). Stereotypes as a measure of success. *Journal of Social Issues, 29,* 83–99.

Thompson, C., Kloph, D., & Ishii, S. (1991). A comparison of social style between Japanese and Americans. *Communication Research Reports, 8,* 165–172.

Tomioko, M. (1994). How Japanese and Americans present themselves in their initial encounters. Unpublished master's thesis, San Francisco State University, San Francisco, CA.

Wilson, K. & Gallois, C. (1993). *Assertion and its social context.* New York: Pergamon Press.

Yum, J. (1988). The impact of Confucianism on interpersonal relationships and communication patterns in East Asia. *Communication Monographs, 55,* 374–388.

Zane, N., Sue, S., Hu, L., & Kwon, J. (1991). Asian-American assertion: a social learning analysis of cultural differences. *Journal of Counseling Psychology, 38,* 63–70.

Part II
Emergence and Impact of Global Media

CHAPTER SIX
Internet as a Town Square in Global Society

Ringo Ma
State University of New York, Fredonia

INTRODUCTION

Gumpert (1998) proposes to modify the context-based classification of communication, including interpersonal communication and mass communication, to issue-based communication studies, such as drug use and budget crises. First, we can not separate one context from another in real life. The interplay of interpersonal and mass communication, for example, can be found in almost every social movement and innovation diffusing process. Second, the level of "massification" associated with mediated communication has been reduced over the past few decades. The large number of available channels with cable television has diversified the viewers who would otherwise be assembled on only a few channels. As Mullins (1996) writes

> Television, as the dominant broadcast medium, is rapidly changing due to technological innovations. Cable television has expanded; the number of specialized channels targeted for micro-audiences is soon to mushroom. Digital compression techniques will allow 500 channels; future fiberoptics innovations will allow even more options. (p. 274)

However, a real challenge to the traditional conception of "mass" communication is the extensive use of networked computers in communication. Communication via the new digital medium has blurred the line between mass and interpersonal communication in a profound way. The Internet enables both personal and public messages to flow across national boundaries more easily and faster than any traditional mass medium. The "massification" created on the Internet is sometimes totally unexpected when a message is sent as a personal message initially. It also provides an opportunity for previously acquainted and unacquainted individuals to communicate from different societies on a regular basis. The two forms of computer-mediated communication (CMC), i.e., personalized mass communication and mediated interpersonal communication, have shattered the boundary between interpersonal communication and mass communication.

Currently computer-mediated communication is presented in various forms though it used to be limited to text transmission:

> Computer-mediated communication starts with text, but, as bandwidth increases, communication will increasingly employ graphic images, which will lead to complex analogous units and structures that go beyond textual paragraphs. There will also be pointers [links on Web pages] to whole bodies of images and information. (Kolb, 1996, p. 19)

From text-based e-mailing to watching and listening to each other's video and audio programs on the World Wide Web, the Internet has changed many traditionally assumed characteristics of communication. First, it has demystified some cultural and societal differences exaggerated through state-sanctioned media programs and yet created a new virtual cosmopolis to facilitate intercultural communication. Second, it is more difficult to gate-keep information and maintain hierarchy in communication, so the voice of the underrepresented is heard more frequently on this new town square. Third, both "hyperself" and "hyperreality" are constructed with the new characteristics of communication and image-creating capability on the Internet. The next section of this chapter will address these three characteristics.

CHARACTERISTICS OF INTERNET COMMUNICATION

Demystification of Exaggerated Cultural Differences

Featherstone (1991) notes that "one of the state's central aims since its formation has been to produce a common culture in which local differences have been homogenized" (p. 147). However, the minimization of within-nation differences can be done at the expense of exaggerated between-nation differences if the agenda behind producing a common culture is to promote patriotism or national pride. National leaders may benefit from the artificially created "we-they" mentality in various ways, such as distracting people's attention from domestic problems and acquiring people's support for their political agenda.

Recognizing cultural differences is an important asset in intercultural communication. Farfetched reports of cultural and societal differences, nevertheless, can make us unable to separate an individual from the cultural and social groups he or she is affiliated with, and thus constitute a major barrier for communication. Many media programs produced in the U.S., for example, exaggerate the differences between Japanese and U.S. cultures. Americans are portrayed as being open and fair to everybody, while Japanese are group-oriented and exclusive in dealing with foreigners. Comparisons can also be made with different time frames to exaggerate cultural differences. An example is to compare foot-binding practice in Ancient China with modern Western thoughts. In addition, cultural and societal differences are sometimes exaggerated through state-sanctioned media programs. When the author was in Chongqing, the People's Republic of China (PRC) in 1991, a newspaper article entitled "Most Americans Lie" notes that, according to a study in the U.S., most Americans admitted they lied at least once recently. No further details were provided about the research, such as how "lying" was defined and in what context the research was conducted. Although the research appeared to be an ordinary study in social science, the way "Americans" are introduced in the article was misleading. It suggests that Americans are liars (while other peoples are not).

Ma (1996) found those who engage in intercultural computer-mediated conversations regularly are better informed about the culture of their communication partners than are those who do not utilize this means of communication. Exaggerated cultural and societal differences are thus eliminated through computer-mediated communication on the Internet. Although censorship is still imposed in some nations, it is more difficult to implement on the Internet than traditional media because there are numerous channels through which messages may be "infiltrated." Because easy access to unbiased information is important in forming a positive attitude toward people from other cultures, the Internet has successfully created a conducive environment for intercultural communication. First-hand information about other cultures is available on the Internet, so we no longer depend on state-sponsored media programs for information. In other words, "international" communication is gradually replaced by "intercultural" communication as CMC occurs more and more frequently than mass communication.

Empowerment of the Underrepresented

Morris and Ogan (1996) note that "the interchangeability of producers and receivers of content" on the Internet makes it different from traditional mass media (p. 44). An audience member can easily become a message producer. This characteristic of the Internet has made the voice of an ordinary person heard much more easily than before. In other words, in addition to breaking through the national line, the Internet has the potential of promoting democracy through empowering people with lower social or economic statuses. As Sproull and Kiesler (1991) note, CMC through the Internet promotes democracy in some organizations:

> In some companies that use computer networking, communication is strikingly open as employees cross barriers of space, time, and social category to share expertise, opinions, and ideas. In a democracy, people believe that everyone should be included on equal terms in communication . . . New communication technology is surprisingly consistent with the Western image of democracy. (p. 13)

CMC enables ordinary people to voice their opinions without going through gatekeepers. This direct participation has altered the traditional pattern of organizational communication. In many large organizations, for example, employees can send e-mail to the chief executive officer (CEO) directly. The CEO is pressured to reply to the message as soon as possible. Otherwise, his or her lack of response to e-mail can easily become a target for attack. On the other hand, employees without access to e-mail tend to have to go through many gatekeepers before their ideas can reach the CEO.

The World Wide Web also allows an employee's voice to be heard more frequently and his or her accomplishments to be judged more fairly. With an inexpensive microcomputer, anyone can be the producer of his or her own presentation package. While gatekeepers can easily filter out written messages transmitted through traditional memorandums, messages presented on the Web are less likely to suffer from suppression. As an easy way to get a picture of what is going on in the organization, the CEO usually visits employees' pages that are linked to the Web site of the organization on a regular basis.

Second, due to the lack of visual and social cues in most situations involving interactions, CMC on the Internet tends to promote egalitarian behavior (Kiesler, Siegel, & MaGuire, 1984). Both East Asian and North American students were found to perceive computer-mediated intercultural communication as a more egalitarian and information-oriented experience than facc-to-face intercultural communication (Ma, 1996). Because it is more difficult to maintain the kind of hierarchy that is present in face-to-face situations, underrepresented people feel more comfortable voicing their opinions on the Internet. Discrimination due to cultural values, physical attributes, and sexual preferences is less likely to occur in computer-mediated intercultural communication.

CMC is counter-hegemonic also because large-scale solidarity among the underrepresented can be easily established. Usually members of underrepresented social groups such as ethnic minorities cannot find many "similar" people in their physical environment. On the Internet, however, they can locate each other easily and exchange ideas regularly. They can also support each other in their virtual community to confront hegemony.

Creation of Hyperself and Hyperreality

Gumpert and Drucker (1992) claim that the traditional notion of physical space co-occupied by communicators is replaced by "electronic space," which is "an associational construct *without place* between two or more persons . . ." (p. 189). The electronic space on the Internet allows previously acquainted and unacquainted individuals who occupy different physical spaces to meet and interact. For example, Liu (1999) describes a Chinese virtual community in North America:

> There, community members enjoy a rich social life. They have access to news in either Chinese or English about their community, China, and the world. They visit Chinese virtual libraries and read Chinese classic literature. They subscribe to a variety of electronic newspapers, magazines, and academic journals in either Chinese or English. They "meet" and "talk" synchronously or asynchronously with their fellow Chinese residing throughout the United States, Canada, and the world. They also have entertainment . . . Via bulletin boards, they offer help, ask for help, and are helped. (p. 195)

Compared with face-to-face communication, audio, visual and other social cues in CMC are, nevertheless, either "lean" (Short, Williams, & Christie, 1976) or can be strategically packaged. Lack of or strategic presentation of these cues has generated a new dimension of communication that is not found in face-to-face communication.

According to Mullins (1996), "hypertext" and "hypermedia" refer to "forms of electronic text that capitalize on the large storage capacity and quick random-access potential of the electronic medium; they are (or can be) integrated and interactive forms" (p. 275). The "hyper" in "hypertext" and "hypermedia," he continues, "really refers to another or additional dimension of text beyond the presently obvious . . ." (p. 275). In CMC, this "new dimension" can as well be extended to the new identity of communicator, "hyperself," and the new reality constructed through CMC, "hyperreality."

Communication involves the interaction among perceived self, desired self, and presenting self. Due to the lack or paucity of audio and visual cues, a new version of the three can be easily developed on the Internet. Unique characteristics of most CMC processes, such as text-based interaction and focusing on the mind permit participants to share their thoughts without being distracted by other social cues. The supporting of each other's presenting self and the perception of this reciprocity are based on a different set of symbolic activities. For example, a poor speaker who can nevertheless write well can develop a much more positive self-concept through online chat than through face-to-face interaction. A girl who is unpopular at her school due to her unattractive physical appearance can also become a popular BBS (bulletin board systems) queen. In other words, the hyperself developed on the Internet can be very different from the version of self in face-to-face situations.

The "hyper" phenomenon has been addressed from different perspectives. Jones (1995) indicates that one who engages in CMC can have multiple identities. Using the Chinese virtual community in North America as an example, Liu (1999) explicates how a member of the community can have multiple identities depending on his or her interests:

> While online, the educator may participate in Id-line (as a Chinese communication scholar), join in soc.culture.china (as a Chinese person), or write poems for chpoem (as a Chinese poet). (p. 196)

In a more negative vein, Nguyen and Alexander (1996) note that in CMC "old assumptions about the nature of identity have quietly vanished" (p. 104). On the surface level, CMC is "highly susceptible to deception" by "identity hackers" (p. 104). An old man can pretend to be a young female, while a White professor can present himself as a Black student. Similar situations have been experienced by many who have participated on IRC (Internet Relay Chat) or BBS talk for an extended period of time. On a much deeper level, CMC can cause the disappearance of a distinct self when numerous online "nicks" are being used:

> Our individual concreteness dissolves in favour of the fluid, the homogeneous and the universal. Once the palpable particularity of individual identity is lost, we become relational feedback units among endless arrays of refracted power. (p. 104)

The CMC version of self-identity portrayed in Nguyen and Alexander's (1996) study is closely related to Jean Baudrillard's notion of hyperreality. Baudrillard (1988) uses the metaphor of a map and its relation to territory to explicate the notion of hyperreality:

> Abstraction today is no longer that of the map . . . Simulation is no longer that of a territory, a referential being or a substance. It is the generation by models of real without origin or reality: a hyperreal. The

> territory no longer precedes the map, nor survives it. . . it is the map that precedes the territory–precession of simulacra . . . (p. 166)

Baudrillard's notion that we live in a "strange new world constructed out of models or simulacra which have no referent or ground in any 'reality' except their own" (Poster, 1988, p. 6) can apply to the virtual reality created on the Internet. A few years ago, a popular U.S. magazine discloses how fantasies may be created by phone sex services. Utilizing special audio signals transmitted through the phone and video images presented in their advertisements, the phone sex companies made their customers believe they were talking to a sexy young woman in a romantic room. As a matter of fact, they might be talking to an unattractive, middle-aged, single mother who was in financial crisis. The hypertext presented on the World Wide Web has the same image-creating capability, but it can reach many more people than phone sex messages. Through a well-designed Web site, a business can project an image quite different from what might be actually seen at its physical site. This "hyperreality" developed from hypertexts and multimedia effects is not only a by-product of, but also is a contextual component of, CMC.

IMPLICATIONS OF THE THREE CHARACTERISTICS OF INTERNET COMMUNICATION

As a new town square of the global village, the Internet has the potential to influence the global society in a profound way. This influence can be grouped into two broad categories: local diversity within international homogenization and the establishment of new world order.

Local Diversity within International Homogenization

The information technology developed to meet the demands on the Internet has made institutional control of information far more unlikely than before. An Internet user in Shanghai can access the same Web site and join the same discussion as another user in New York City. Through streaming audio and video programs we no longer depend on the short wave radio or satellite receiver for international media programs. For example, through a streaming video player people around the world can watch the same TV programs as those in Taiwan.

According to Parsons (1966, 1971), a social system has the following four subsystems: economic, political, social and cultural. Robertson (1992) contends that globalization, which is defined as "compression of the world" and "global consciousness," has occurred at all the four levels. Hoogvelt (1997) describes how the world has shrunk to a "global village":

> Today's telecommunications using satellite TV and the linking of computers through cyberspace allow most 'disembodied' services, for example technological designs, managerial instructions, and operational controls, as well as media images of wars and earthquakes and representations of consumer fashions, to enter the minds of people instantly anywhere in the global system. (p. 120)

Based on Robertson's conception of globalization, Hoogvelt (1997) also explicates how global consciousness is intensified through world compression:

> Global consciousness is manifested in the way we, peoples all over the world, in a discourse unified through mass communication, speak of military-political issues in terms of 'world order' or of economic issues as in 'international recession.' We speak of 'world peace' and 'human rights,' while issues of pollution and purification are talked about in terms of 'saving the planet.' (p. 117)

With both the shrinking of physical space and sharing of minds, the building of a global social system is in progress. More specifically, the economic, political, social, and cultural subsystems of a global society are forming. In this process of globalization, the Internet becomes the most powerful among all means of telecommunication, because it has the potential to replace others in the near future. Many of us are gradually substituting the Internet for traditional TV, radio, telephone, and facsimile as means of communication.

In a sense, the Internet serves as a town square for direct access among people at different geographic locations. Although most of those who meet on the Internet may not be as confrontational as in a traditional town square, a much wider repertoire of perspectives can be generated in this global town square. Furthermore, given the fact that there is no universal standard for resolving conflict, the interaction *per se* can become a promising form of mutual influence.

On the other hand, the voices of the traditionally underrepresented are heard more frequently than before. They can unite and exchange ideas across national boundaries through the Internet. For example, many social groups for environmental protection and women's rights are global in nature. Aboriginals from different nations are also beginning to meet one another. They share similar concerns and work together to protect their rights. As more and more people have access to the Internet, the use of mass media as the mouthpiece of the ruling party is unlikely to last very long.

Both cultural domination and cultural diversity have been proposed as the consequences of modern technology. Featherstone (1991), however, argues that globalization leads to a widening range of different cultural contacts with others. Although the world is becoming a singular place, each local culture will not be a unified one. In other words, different cultures in the world will maintain close contact with one another and have the opportunity to influence one another, while within each local society cultural diversity and multiple identities will be established.

One way to compare the world in the past and in the future is to use the analysis of variance in statistical analysis as an analogy. In the analysis of variance procedure, the "between-sample variability" is compared with the "within-sample variability." If the formal gets smaller and/or the latter gets larger, it will be less likely to reject the null hypothesis that there is no difference between two groups. The new global society that is in the forming is similar to a low "between-sample variability" associated with a high "within-sample variability." It will be less easy to identify cultural differences between nations. On the other hand, it will also become more difficult to define what a particular national culture is.

Development of New World Order

The traditional way of maintaining social order is largely based on the assumption that two persons have to be physically close to cause a conflict in interests. The identity and reality (re)construction that occurs on the Internet has nevertheless challenged this assumption. On the Internet, for example, a purchase order for a locally illegal item can be placed in a different nation. Offensive language that is sometimes interpreted based on different social standards can be exchanged between Internet users from two different nations. Children have easy access to a large amount of pornography supplied from all over the world. The social control force for such activities is either as formal as the constitution of a nation ("legality") or as informal as the pressure from peers ("shame"). Although they have all happened before, it was generally much more difficult to pass the legal boundary in a traditional or pre-Internet society.

In many ways, computer-mediated communication on the Internet nurtures irresponsible behaviors such as lying and fabricating. The global town square on the Internet can easily become a lawless and shameless public domain, in which no one is concerned about the well-being of others. The current social and legal system is not sufficient to handle these new problems.

Therefore, a new world order needs to be established before we can truly enjoy the fruits of modern technology. Among many possible measures that we can take, a global identity verification system can be implemented to prevent crime on the Internet. The system should not only prevent universally unacceptable conduct but also protect the privacy of users. To implement the system, the technological requirements may not be as challenging as minimizing unnecessary political influences. Because each nation is likely to argue for a different set of rules governing the verifying procedure in favor of their own national policy, it might be very difficult to reach a universal agreement.

Special rules are also needed to ensure universally acceptable behaviors and procedures for participation in discussion through CMC. Whenever there is an issue to be discussed on the Internet, the traditional parliamentary procedure seems to be insufficient to handle the virtual gathering. Therefore, a detailed on-line parliamentary procedure will be required if many decisions are to be made through the Internet in the next century. Currently informal guidelines called "netiquette" (network etiquette) have been formed for some bulletin board programs. For example, Usenet has the following:

> 'Proper' netiquette dictates that posting should be distributed in as limited a manner as possible; that the same article should not be posted twice to different groups; that posting should not be repeated; that posting other people's work without permission is not allowed; that postings should be appropriate to Usenet and the newsgroup; that postings should not be used for blatantly commercial purposes. (Shade, 1996, pp. 13–14)

However, the guidelines are for posting messages only. Detailed procedures are also needed for synchronous idea exchange and decision making on the Internet.

Some macro issues have to be addressed as well. Cultural imperialism imposed by technologically advanced nations has been a major concern of many scholars. As Hoogvelt (1997) writes:

> We need to conceptualize the emerging governance by the global capitalist class as a complex process that institutionalizes structural power through the wide-spread adoption of cultural values and legitimating ideology. But this legitimating ideology, while parading under the banner of 'deregulation' and 'privatization', yet draws in governments in an ever-widening circumference of 'regulation' in the form of policy initiatives and legislation. These include monetary and

> fiscal policies . . . and even the reconstitution of social obligations; for example, an ideological attack on alternative lifestyles, and prioritization of traditional family values through social policy initiatives. (p. 138)

Protection of those cultures that tend to avoid technological advancement and yet have a significant spiritual heritage to share with the rest of the world has become a legitimate concern. It is important to find a way to protect their voices on the Internet as well as their cultural heritage. McChesney (1996) notes that "the structural basis for genuine democratic communication lies with a media system free from the control of either the dominant political or economic powers" (p. 108). Thus, to make the Internet an authentic town square, regulations have to be established to keep it free from undesirable control. Based on Jürgen Habermas's critical theory, Ess (1996) proposes the following minimal conditions for communication in the new community:

> The discourse ethic requires the ability to engage in critical discourse and the moral commitment to practicing the ability to take others' perspectives and thus seek solidarity with others in a plurality of democratic discourse communities. Such requirements are neither unusual nor remarkable: On the contrary, they seem to be minimal requirements for citizenship in a democratic society . . . (p. 220)

In other words, the new world created by the Internet is different from the previous one in many ways. In order to prevent some detrimental effects of the Internet from taking place, a new world order has to be established. The new world order should aim at maintaining mutually acceptable behaviors and being receptive to underrepresented voices.

CONCLUDING REMARKS

CMC on the Internet has created a global digital village that was never realized through traditional mass media. It is a communication-intensive town square, rather than merely an arena for elite media programs. On this town square, the line between interpersonal communication and mass communication is blurred. Villagers experience both personalized mass communication and mediated interpersonal communication. They not only recognize cultural differences but also enjoy the freedom from being brainwashed by state-sanctioned media programs. The town square has fulfilled Featherstone's (1991) postmodern prospect of "unity through diversity":

> The abandonment of such state-led cultural crusades and nationalist assimilation projects which were central to modernity is one symptom of the move towards postmodernity. (p. 147)

The digital town square is counter-hegemonic in many ways. The voice of the underrepresented is heard more frequently than before because it is more difficult to gate-keep information and maintain hierarchy. Anyone can produce his or her public relations programs with inexpensive microcomputers. It is also easier than before for the members of various muted groups to unite and confront dominant groups.

The self-image projected and the reality constructed through CMC demonstrate a new extension of our physical being and environment. The hyperself can be very different from its face-to-face counterpart, just as the images projected on TV can differ significantly from those on radio. The hyperreality is a joint product of interactions among various hyperselves in the global digital village. Proper management of our hyperself in order to create a hyperreality that we like to see is becoming more and more important in this increasingly digital new world.

The three characteristics of communication developed through the use of the Internet imply the emergence of an internationally homogeneous and yet locally diverse society in the next century. They also warrant the establishment of a new world order as a way to prevent negative impacts from the Internet.

A recent survey finds that eight to 12 million workers in the United States are logging on ("telecommuting") rather than commuting, and that "half to two-thirds of American companies permit some employees to telecommute" (Stafford, 1998). Although both employers and telecommuting employees raised many issues regarding the new organizational phenomenon, "observers say telecommuting is growing at 20 percent annually" (not including the self-employed) (Stafford, 1998). Therefore, the global digital village is by no means at an illusion stage. It *is* in the forming. At least we are technologically ready for this change. However, as communication scholars, we need to ask the following questions: Are we conceptually prepared for the new form of communication; and in higher education, have we prepared our students for the imminent global town square, or are we simply emphasizing what has been done in the past?

REFERENCES

Baudrillard, J. (1988). *Simulacra and simulations* (P. Foss, P. Patton, & P. Beitchman, Trans.). In M. poster (Ed.), *Jean Baudrillard: Selected writings* (pp. 166–184). Stanford, CA: Stanford University Press. (Reprinted from *Simulacra and simulations*, pp. 1–13, 23–49, by P. Foss, P. Patton, & P. Beitchman, Eds. and Trans., 1983, New York: Sémiotext)

Blumer, H. (1969). Symbolic interactionism: Perspective and method. Englewood Cliffs, NJ: Prentice Hall.

Ess, C. (1996). The political computer: Democracy, CMC, and Habermas. In C. Ess (Ed.), *Philosophical perspectives on computer-mediated communication* (pp. 197–230). Albany, NY: State University of New York Press.

Featherstone, M. (1991). *Consumer culture and postmodernism.* Newbury Park, CA: Sage.

Gumpert, G. (1998, April). *The rise and fall of mass communication as a discipline.* Paper presented at the 1997 Convention of the Eastern Communication Association, Saratoga Springs, NY.

Gumpert, G., & Drucker, S. J. (1992). From the Agora to the electronic shopping mall. *Critical Studies in Mass Communication, 9,* 186–200.

Hoogvelt, A. (1997). *Globalization and the postcolonial world.* Baltimore, MD: Johns Hopkins University Press.

Kiesler, S., Siegel, J., & McGuire, T. W. (1984). Social psychological aspects of computer-mediated communication. *American Psychologist, 39,* 1123–1134.

Kolb, D. (1996). Discourse across links. In C. Ess (Ed.), *Philosophical perspectives on computer-mediated communication* (pp. 15–26). Albany, NY: State University of New York Press.

Liu, D. (1999). The Internet as a mode of civic discourse: The Chinese virtual community in North America. In R. Kluver, & J. H. Powers (Eds.), *Civic discourse, civil society, and Chinese communities* (pp. 195-206). Stamford, CT: Ablex.

Ma, R. (1996). Computer-mediated conversations as a new dimension of intercultural communication between East Asian and North American college students. In S. C. Herring (Ed.), *Computer-mediated communication: Linguistic, social, and cross-cultural perspectives* (pp. 173–185). Amsterdam, Netherlands: John Benjamins.

McChesney, R. W. (1996). The Internet and U.S. communication policy-making in historical and critical perspective. *Journal of Communication, 46*(1), 98–124.

Morris, M., & Ogan, C. (1996). The Internet as mass medium. *Journal of Communication, 46*(1), 39–50.

Mullins, P. (1996). Sacred text in the sea of texts: The Bible in North American electronic culture. In C. Ess (Ed.), *Philosophical perspectives on computer-mediated communication* (pp. 271–302). Albany, NY: State University of New York Press.

Nguyen, D. T., & Alexander, J. (1996). The coming of cyberspacetime and the end of the polity. In R. Shields (Ed.), *Cultures and internet* (pp. 99–124). Thousand Oaks, CA: Sage.

Parsons, T. (1966). *Societies*. Englewood Cliffs, NJ: Prentice Hall.

——. (1971). *The system of modern societies*. Englewood Cliffs, NJ: Prentice Hall.

Poster, M. (Ed.). (1988). *Jean Baudrillard: Selected writings*. Stanford, CA: Stanford University Press.

Robertson, R. (1992). *Globalization*. London: Sage.

Shade, L. R. (1996). Is there free speech on the net? Censorship in the global information infrastructure. In R. Shields (Ed.), *Cultures and internet* (pp. 11–32). Thousand Oaks, CA: Sage.

Short. J., Williams, E., & Christie, B. (1976). *The social psychology of telecommunication*. London: Wiley.

Sproull, L, & Kiesler, S. (1991). *Connections: New ways of working in the networked organization*. Cambridge, MA: MIT Press.

Stafford, D. (1998, May 25). Telecommuting Just Keeps Rolling. *Saint Paul Pioneer Press*, p. 6D.

CHAPTER SEVEN

"Village Work": An Activity-Theoretical Perspective Toward Global Community on the Internet

G. Richard Holt
University of Illinois at Chicago

INTRODUCTION

Like so many technological advancements before it, the Internet has in recent years been extravagantly promoted (particularly in popular press and advertising) as a means whereby all humanity can be brought together. Optimistic predictions abound of a "global community," linked by instantaneous transfer of information, crossing boundaries of nation and language, and comparatively free of the regulatory constraints of individual nation-states. While some of the promise of forging computer-mediated communities has certainly been realized, the goal of large-scale collectivities linked by computer has remained tantalizingly beyond our grasp. There could be a number of reasons for this: lack of availability of computers to many of the world's people; inadequate infrastructure in developing nations to support electronic transfer of information; illiteracy; fear of excessive computerization of society; and so on.

However, there may be another reason. The current technologically advanced level of computer-mediated communication (CMC), coupled with considerably less advanced levels of cultural integration, have combined to produce an inevitable need state which makes it certain there will be a struggle to move past the contradictions that now prevent us from realizing the full potential of CMC to form global communities. We are at a crossroads where we are confronted with what Karl Marx referred to as the paradox of production and consumption: to consume, one must produce; to produce, one must consume. In terms of CMC in the global society, this means that production (in the form of ever-accelerating technological advancement) gives rise to greater levels of knowledge about others with whom we seek community; however, it also means that absorption of such knowledge (consumption) depends upon ever-increasing levels of its production (both in terms of volume and sophistication). To put it another way: the better our technology gets at disseminating information, the more critical becomes the need to put that information to use in some socially beneficial way.

In this chapter, the production-consumption paradox of communities bound by CMC is considered through the lens of activity theory, as proposed by the Finnish organizational theorist Yrjo Engestrom (1987). Engestrom's unified field analysis of the production-consumption paradox, which he and others have used to successfully tackle a broad range of organizational problems (e.g., Engestrom, 1987, 1990; Engestrom & Middleton, 1996), offers opportunities for scholars to examine afresh the problems of forging CMC communities, particularly those which (like the elusive "global society") present challenging problems involving the synthesis of dramatically divergent cultural traditions.

Following a brief introduction to activity theory, I will examine the activity systems and subsystems involved in a year-long collection of postings to a public bulletin board, part of an ongoing worldwide teaching/learning project which high school students use to discuss utopian communities.

ACTIVITY THEORY

What Is Activity Theory?

Activity theory began with Lev Vygotsky (1978), whose work was the foundation of what is now often referred to as the cultural-historical school of Soviet psychology. One of Vygotsky's projects was to come up with an alternative to the implicit division of mind and behavior that had become so much a part of Western psychological thinking. To do this, he proposed a type of psychological knowledge based on integrating, rather than separating, the individual and social environment. Cole (1985) summarized Vygostky's position:

> Central to [Vygotsky's] effort was an approach that denied the strict separation of the individual and its social environment. Instead, the individual and the social were conceived as mutually constitutive elements of a single, interacting system; cognitive development was treated as a process of acquiring culture. (p. 148)

Rather than follow the implications in Western psychology of the Cartesian dualistic separation of mind and world, the scholars of the cultural-historical school (e.g., Bruner, 1985; Leontyev, 1978; Luria,

1976) proposed instead that the interaction of minds and worlds exists a priori (Bakhurst, 1988). This position has a number of important theoretical implications, perhaps the most significant of which is that "the higher mental functions are internalized forms of the activity of the community in which an individual acts" (Holt & Morris, 1993, p. 98).

One key outcome of the work of the cultural-historical school has been to refocus attention away from generalistic theories about social life, and toward consideration of each life experience as a sociohistorically specific act. The cultural-historical school has been relentless in its efforts to redirect psychology away from its emphases on ideas abstracted from sociohistorically specific behaviors, and toward considering each human activity as uniquely resulting from internalizing higher mental processes acquired through culture.

One of the most interesting and provocative expansions of Vygotsky's ideas is in the work of Yrjo Engestrom (1987), who presented a unified model of how human collectives work, based principally on two ideas: (1) an expansion of Vygotsky's theory linking subject, object, and mediating instrument, to include elements of the collectivity (rules, community, and division of labor); and (2) a rigorous conceptualization and linking of various elements of what Marx (1973) referred to as the "paradox of production and consumption." Engestrom derived an extraordinarily elegant yet challenging model of how humans work in collectives, which he called "the triangle of activity" (see Figure 1).

To explain this diagram, comprised of a large triangle, four subtriangles ("production," "consumption," "exchange," and "distribution"), and six nodes (subject, instrument, object, rules, community, and division of labor), let us begin with the top, a smaller triangle inscribed by the subject, instrument, and object nodes. As Holt and Morris (1993) described this top triangle:

> The "subject" is either an individual or aggregate of individuals seeking to fulfill goals or motives through action (if individual) or activity (if groups). "Instruments" can be defined as means (concepts, theories, physical apparatuses, logical reasoning, to name only a few that Engestrom has employed) that mediate the subject's activity toward the object. "Objects" are modifiable ends toward which activity is directed and from which an outcome is expected. (p. 98)

The top smaller triangle (called the "production subtriangle") is an approximate graphical summary of Vygotsky's idea of mediation that signs acquire meaning through a series of qualitative transformations and arise from something that is not originally a sign operation. In similar fashion, human activity, as Engestrom conceived of it, consists of human

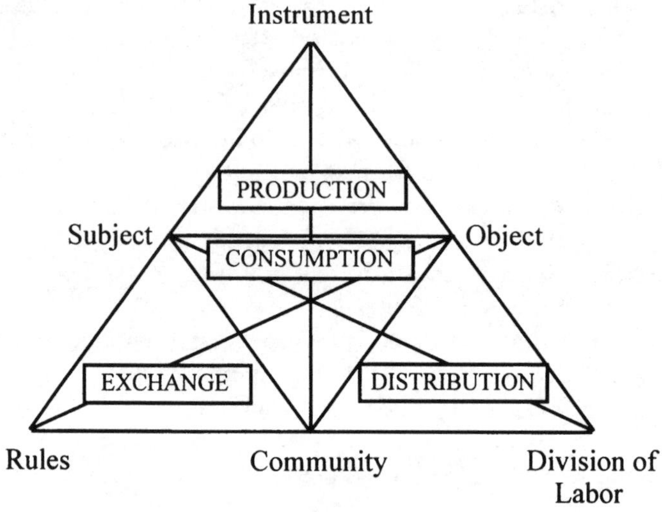

Figure 1
Triangle of Activity
(Engestrom, 1987)

beings using some mediating means (instruments) in the service of achieving a particular end (object), from which is expected an outcome.

Looking at the overall activity triangle, however, one also sees how the work of Vygotsky stopped at the individual level, leaving the idea of the social world—in which every individual must of course operate—relatively undeveloped and implicit. It remained for Engestrom to "flesh out" the links between the individual and the broader social environment in which activity is performed, and thereby to achieve a situated account of activity. To fashion this turn, Engestrom used an idea from Marx (1973): human work is founded on the notion that, although work, considered as an overall activity, is geared to production, four elements involved in the work activity are designed both to produce and consume—these elements Marx referred to as "production," "consumption," "exchange," and "distribution." He summarized their interrelationship as follows:

> Production creates the objects which correspond to the given needs; distribution divides them up according to social laws; exchange further parcels out the already divided shares in accord with individual needs; and finally, in consumption, the product steps outside the social movement and become a direct object and servant of individual need, and satisfies it in being consumed. (Marx, 1973, p. 89)

Hence, the paradox faced by all humans who must perform activity in collectives is that production is not simply production—production is also consumption. This, as Holt and Morris (1993) note, "is the motive force driving all activity systems: were it not for the paradox that consumption necessitates production, and vice versa, activity would not exist" (p. 99).

Graphically, Engestrom represents the subelements of activity by having the larger triangle (representing the whole activity system) encompass four smaller triangles, the first labeled in Figure 1 as "production" (this is the subtriangle previously identified as roughly analogous to Vygotsky's notion of mediation). With the addition to the large triangle of three more nodes—rules, community, and division of labor—Engestrom is able to derive three other subtriangles, and he names these as Marx did: "consumption," "exchange," and "distribution." It is in the four subtriangles (representing smaller activity subsystems) that the core of the production/consumption paradox actually resides—it is in the interactions among the elements represented by the nodes of the subsystems that consumption (along with production) primarily occurs, thereby placing the smaller systems in dynamic tension with the overall activity system, which is geared primarily toward production. Thus, any group of individuals initiating any activity automatically engender these forces and counterforces on the way to achieving the object and, from there, the outcome. As we will see, this formulation is particularly applicable to messages posted to bulletin boards on the Internet, where the initiation of activity (i.e., posting) is usually performed with little or no awareness by the poster of any immediate, personal response the post might engender.

Anyone who attempts to describe human activity by applying Engestrom's graphical representation quickly comes to realize the depth and complexity of his theory. It would be well to caution the reader that, like other proponents of the cultural-historical approach, Engestrom is little concerned with graphic (or any other) representation of activity, per se. The aim of activity theorists is to use the graphical representation (often achieved, as we will see, at extraordinary cost in mental effort, time, and research) as a "way in" to transforming in the organization which the activity system represents. What Engestrom looks for are contradictions, arising from the system's production-consumption paradox, which give rise to what he calls "need states," from which emerge new instruments (these Engestrom calls "springboards," " ... new specific instrument[s] ... for breaking the constraints ... and for constructing a new general model for the subsequent activity" [Engestrom, 1987, p. 189]). An important insight from activity theory is that need states, and their resulting springboards, are not only possible, nor merely likely, but inevitable. In a way, the subject who wishes to produce by means of the activity system becomes, by his or her activity, locked into the four subsystems (through the connection of the subject node to the rules, object, and community nodes), therefore becoming unable to avoid the production/consumption paradox.

An Example of an Activity Theoretical Reading

To illustrate, let me trace what, in terms of activity theory, might be happening to a typical message posted to a bulletin board (BBS). (This example is purely hypothetical and does not relate to the case study later analyzed in this chapter.) The subject is a group of posters, who intend to achieve the object of having their message read (and perhaps responded to), and who use—as an instrument—the technology associated with the Internet. (For simplicity's sake, I will use "Internet" as a covering term for all the various elements [instruments] involved in Internet communication—computers, application programs, servers, and so on.) Let us further specify that the message is a certain type: an announcement about an upcoming television show the posters (subject) think might be of interest to others on the BBS. As an outcome for this object, we could say the posters want those who read their message to see the program (and, perhaps, ultimately to share insights about it with other BBS members).

At this point, the whole situation seems relatively straightforward. However, as we know from Engestrom's expansion of Vygotsky's account of mediation (limited to the same basic ideas as the top, or "production," subtriangle), any activity necessarily involves others. Hence, inevitably, there are brought into the system rules (for simplicity, let's say the rules relate to what type of messages are appropriate for the BBS), sanctioned by the community (i.e., the BBS membership), and apportioned through division of labor. (For example, a message about an upcoming television show may not be posted by just anyone, but only those who, [a] know about the program, and [b] care enough to tell others about it.) The division of labor node could also describe the specialty activities engaged in by other BBS members. For example, one could envision (at least) the following kinds of audience "specialists": members who hear of the program, but supply different broadcast times from their own local markets; members who have seen this particular, or a similar, program in the past and wish to comment; members who see the list as a refuge where they can avoid discussing things like television and wish to register an objection; and so on. Based on their unique experiences, all respondents have the potential to formulate idiosyncratic responses to the posters' announcements, and hence can all be considered "specialists" performing according to their own ideas about division of labor.

However, beneath the seeming calm and even banality of this common e-mail transaction lurk forces of conflict, awakened by the "simple" activity of posting. In each of the four smaller activity subsystems (subtriangles), there is a struggle between production and consumption. In the production subsystem, for example, we cannot say the subject proceeds to the object at no cost: we all know how much using the Internet (the instrument node) costs, not just financial costs associated with logging on to the server, but economic costs of computer systems, and even cost in time and effort involved in composing the message itself. To produce the message, then, the subject must consume resources, some attached to the instrument, some not.

In the consumption subsystem, there is also evidence of the production/consumption dynamic. If there is no community to perceive the message and give it meaning, what is the worth of posting it? Behind the production of a message is often the expectation that there is at least a potential audience who will read it and respond to it. However, the effort needed for the audience (the community node) to respond is much the same as it is for the original poster; once a message is posted, someone is likely to make an effort to read it and may even respond. Among the potential readers will of course be teachers evaluating the poster, as a student. All these actions involve consumption of the resources of the community.

Likewise, turning to the exchange subsystem, we see that the rules produced by the community—the guidelines according to which the subject interacts with others on the BBS—are achieved at cost to the activity system in general, and the exchange subsystem in particular. BBS maintainers, together with those who are "merely members" of the BBS, must design and test computer systems; monitor activity; and formulate/apply standards of online behavior for BBS participants. For the subject to produce an exchange with the community, the exchange subsystem must consume the efforts of site maintainers and others.

Finally, one also finds in the distribution subsystem undercurrents reflecting the push-and-pull of production and consumption. To achieve the object (the student's posted message), there must be specialists in the community, operating according to a division of labor. For example, one often finds on a given BBS someone performing the specialized role of moderator. On any BBS, one also finds "old hands," members who have been with the BBS longer—these people can (at times) be easily distinguished from the recently-joined, or "newbies." These and other specialists, by using their unique skills, abilities, experience, and knowledge, are often able to frame a message, either directly or indirectly, so that it either does or does not create results its author might have intended. Again, however, the behind-the-scenes expenditure of effort by these specialists confirms that distribution is a product attained at the cost of consuming such commodities as time, money, and mental thought.

Taking all this into account, when we look "backstage" at the activity which drives the "simple" post of an announcement to a BBS, it is clear things are not at all simple. Indeed, in each of the four subsystems reside complex interlinkages of production and consumption. Moreover, it is worth noting that the analysis of this hypothetical activity system has only scratched the barest surface of all the various elements; were one to try to uncover all the potential variables in the act of posting, one would find it a very daunting task.

It should also be noted that this analysis is a pencil sketch of the full-color oil portrait of the activity system that might be rendered later. Indeed, even the full explanation of an activity system (the better part of which, for a sample case, will be advanced in the data analysis portion of this chapter) is merely preparatory to using that description as a way of bringing to light contradictions in the real, living collectivity, so that adjustments can be made, and their results observed—activity analysis is an ongoing, never-ending process, because human collectivities are themselves ongoing and never-ending. If the activity theorist were to take refuge in the reification of an activity system in the form of an "activity triangle" (or any other representation), s/he would commit the same offense the cultural-historical school laid at the door of Western psychology: coming to rely on abstractions, rather than keeping focused on living activity systems. As we turn to the activity analysis in our case study, remember that the activity theorist would regard these findings as a starting point, not as an end in themselves.

ACTIVITY IN INTERNET DISCUSSION: A CASE STUDY

Description of the DORADO Group

Established in 1995, DORADO (not its real name) is a collaborative learning project which allows secondary school students, aged 14 to 19, to use Internet communication to reflect on themes of utopianism and dystopianism, as well as the place of their own and other cultures in the continuum of history. Each year, a varying number of secondary schools from all over the world participate in DORADO projects; for example, in 1998, the year which provides the data sample for this case study, eight schools participated, from the following countries: Belarus, Estonia, Ghana, Lithuania, Mexico, Romania, South Africa, and the United States. DORADO encompasses a number of learning activities, including the following: (1) a discussion group in which students post to a BBS on issues related to quality of life over an approximately three-hundred-year period (1898 to 2098); (2) an archive of student-composed reports on various topics relevant to utopian and dystopian thought; (3) an archive of discussions between project coordinators (on a BBS different from the one students post to); (4) a forum for discussing utopian topics; and (5) a reference list of published work on utopianism and dystopianism. All these materials are online and available for public scrutiny.

To look for indications of activity systems and subsystems at work in DORADO's discussions, I examined messages posted to the BBS for 1998; there were a total of sixty-nine messages, arranged in twelve message "threads" (topic-based sequences of posts-and-responses). A summary of the message threads, together with my estimate of their relevance to the topic of utopianism (according to a five-category Likert-type scale, ranging from "very high" to "very low") is presented in Table 1.

As might be imagined, the ideas expressed in many of the postings to the BBS seem highly culture-specific. Moreover, although high school classes from eight countries were listed as participants, students from five countries (Belarus, Ghana, Lithuania, Romania, and South Africa) do not appear to have participated at all on the BBS (at least, there are no overt cues in the posted messages indicating writers as being from these countries), even though some these students did participate in other site activities, such as essay writing. The greatest volume of discussion by far occurs between students and teachers from Estonia and the United States, with a few contributions by three students from Mexico. Apparently, even when the topic is utopian society, full participation is hard to achieve.

Identifying the Target Activity System

There are a multitude of activity systems that could be analyzed from the 1998 DORADO message board. Since in this chapter it would be possible to deal adequately with only one system, my first step was seek an activity system central to the operation of the DORADO group, yet one whose examination would also shed some light on the question of achieving global community through the Internet.

After examining the body of messages, it struck me that one problem emerged as greater in importance than all others: how organizers and maintainers of the BBS (teachers all) were compelled to keep reorienting discussion to topics which reflected the pedagogical goals of the

Table 1
MESSAGE THREADS
POSTED TO DORADO BBS, 1998

Thread No.	Topic	No. of Posts	Relevance
1	Mission statement for humanity	5	Very high
2	Television and culture	20	Neither high nor low
3	State of the Union	2	Neither high nor low
4	Dreams and responsibility	1	High
5	Adding reports	2	Very low
6	Young people and music	9	Low
7	Sports	2	Very low
8	Military and utopia	4	High
9	Do we need the Internet?	15	Neither high nor low
10	Is democracy good?	6	High
11	Learning languages	2	Neither high nor low
12	Symbolism	1	Low

DORADO group, as against the tendency of DORADO members to assert beliefs to which they had been acculturated, in their native cultures, all their lives. Because DORADO deals with teaching, I chose to frame the opposing forces in the activity system as formal and informal education. What is posted on the DORADO board accords to formal standards because it is part of customary classroom procedure; at the same time, it accords to informal standards, in that all students rely on lessons they have learned from everyone who has taught them in the various cultures to which they have been assimilated.

Not only is the problem of formal versus informal education part of nearly all the message threads, but, since the majority of postings were from students in Estonia (a former state of the USSR) and the United States, discussions brought to light differing views between students of these two countries regarding such topics as individual freedom (Estonia gained its independence from the USSR in 1991, whereas the United States has been a democratic state for more than two centuries). This and similar distinctions that will become evident as we analyze DORADO's activity system go to the heart of the question of whether it is possible to achieve global collectivities—indeed, it is in one way reframing the persistent question about globalization: can people in disparate cultures put aside difference in the service of forging unity?

Performing the Activity Analysis

Having decided to focus on the activity system reflecting the tension between pedagogical and acculturated education, I followed the four steps of activity analysis proposed by Holt and Morris (1993): (1) define the nodes of the activity triangle; (2) discuss the production/consumption paradox inherent in the system; (3) identify the primary contradictions arising at each of the activity triangle's nodes; and (4) demonstrate the emergence of secondary contradictions from the primary contradictions (p. 104). The preliminary description of what I will call the "formal/informal teaching activity system" can serve as a starting point which could eventually be used to move the system through a series of qualitative transformations that would make participants and maintainers more aware that need states, contradictions, and springboards are inevitable and natural aspects of life on the DORADO BBS. However, even though the initial derivation of the activity system is based upon careful and thorough analysis of the message corpus, its efficacy as a transformative mechanism can only be determined, (1) in application to the actual, living, sociohistorically specific system, and (2) in the realization that the initial formulation not only can, but must, be changed. In other words, the activity system formulation is provisional, in that it awaits development of the factors associated with the evolution of the activity system itself before it is possible to determine whether identification of system elements (and their interaction) truly reflects the state of the system. Keep in mind, too, that this preliminary formulation proceeds only to the point of identifying potential secondary contradictions. In an actual activity theoretical analysis, tertiary and quaternary contradictions would be addressed as well. These caveats in mind, we now turn to the four steps followed in analyzing the target system.

Four Steps in Analyzing the DORADO Formal/Informal Teaching Activity System

Step 1: Define nodes of activity triangle. The goal of this initial formulation is to more thoroughly define and explicate the seeming contradiction arising from teachers' attempts to get students to perform according to the standards they have laid out for academic performance, as against the tendency of students to express themselves in terms that relate, not (necessarily) to the goals of the teachers, but to knowledge students have gained through acculturation in their own "home" cultures. Obviously, the subject node must be "teachers," the instrument node is "the Internet," and the object node is "learning," with its associated outcome of "concept mastery." It should be kept in mind that designation of teachers to occupy the subject node is based on the decision to define the activity system as related to DORADO teaching procedures; this should not be taken to imply that the student is not an active participant. It would be useful—indeed, essential—for a complete activity theoretical analysis to be cognizant of another activity system, one in which the student would be the subject. Activity theorists would then find it useful to compare what are called "quaternary contradictions," that is, contradictions between one activity system and another neighboring system. To do a thorough activity theoretical analysis, one would need to identify and inter-relate a number of activity systems.

Moving to the lower part of the formal/informal activity triangle, things become a bit murkier. Conceivably, "rules" involved in expressing oneself to indicate concept mastery could include virtually anything in the student's cognitive repertoire, from academic knowledge to familiarity with American pop culture and slang. However, in this analysis, we are focusing on the intent of the teachers to produce an overt outcome in students' behavior (in the messages students compose and post) that may or may not be at variance from behaviors to which they have been acculturated; therefore, the rules node should represent "rules of culture," that is, rules which govern how one expresses oneself. Again, it will be more convenient to regard "culture" as a covering term, ignoring (for the moment) the manifold elements that comprise cultural rules of self-expression.

The community node is comprised of all those with an interest in the formulation of the student's posting (and therefore the outcome—concept mastery). While this could conceivably include every person the student has ever had contact with, we are primarily concerned with those in the student's life who have the most influence on his/her ability to formulate written posts and responses—not only DORADO teachers, but also the primary acculturators (parents, friends, significant mentors) in the student's home cultures. These groups will be designated by the covering term, "educators" (with the understanding that "education" occurs both in and out of formal classroom settings).

The only remaining node is division of labor. Here it seems appropriate to delineate the specialized jobs of the various people who are, or have been, responsible for the way in which the student expresses him- or herself. Obviously, many people (including not only DORADO teachers and administrators, but others who teach, or have in the past taught, the student) have influenced how students think and write. Likewise, the peers of students often perform specialized tasks modifying the way students express themselves (peers, in fact, frequently exercise on teenagers a far more powerful influence than formal educators). Additional significant others in the student's "home" cultures (parents, friends, mentors, and so on) perform the specialized function of providing role models the student can emulate as s/he thinks and writes. Those in mass media (including the Internet) also provide templates for students to express themselves on the BBS. The covering term for this node will be "specialized tasks related to acculturation."

It might be wondered how we can know that the nodes defined above accurately and truly represent the activity system we want to analyze. For example, what would prevent us from delineating one of the nodes—say the community node—as "Internet technicians," rather than, "educators"? The answer is that, if an informed analyst concludes that one of the nodes should represent something else in the activity system, then that provisional interpretation can also be accepted as legitimate. Recall that the validity of the derived activity system can only be confirmed in application to the actual workings of that system. In other words, for the formal/informal teaching system, the question of whether the community node should be "educators" or "Internet technicians" (or something else entirely) can only be validated by experimentation with the activity system as it evolves. In its approach to ongoing revision of theoretical perspective, activity theory is similar to Glaser and Strauss's (1967) "grounded theory" approach, an inductive method in which researchers begin with observations and then propose categories, themes, or patterns, aware they might have to accommodate unexpected findings later on. The formal/informal education activity is illustrated in Figure 2.

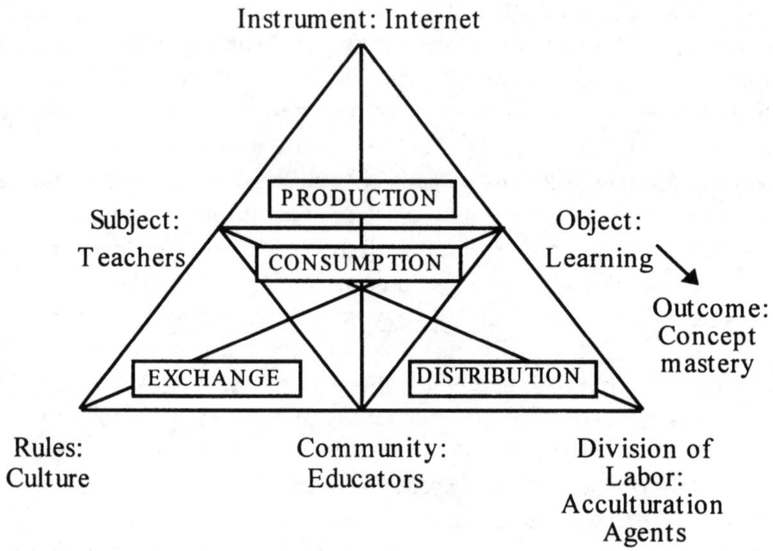

Figure 2
Components of Activity Triangle Representing Formal/Informal Teaching System, DORADO Group

Step 2: Discuss production/consumption paradox inherent in system. Recalling the production/consumption paradox—that the overall activity system is geared to production, while the production, consumption, exchange, and distribution subsystems are geared to both production and consumption—it is now necessary to identify, (1) the principal product of the overall activity system, and (2) the production and consumption taking place in each of the four subsystems.

The overall system, as defined, is geared to produce the object of learning and the associated outcome of concept mastery. To determine what other kinds of production, as well as consumption, are active in this system, I will analyze the four subsystems, one at a time.

In the production subsystem (see Figure 2, subtriangle circumscribed by subject <-> instrument <-> object/outcome), production has principally to do with direct teaching, teacher to student, via the Internet.

In DORADO BBS messages, one sees numerous examples of teachers instructing, admonishing, correcting, encouraging, and cajoling students regarding the topics under discussion. One of the DORADO organizers, Mr. Winston (note: all names and net-names have been changed to be fictitious), for example, confronts a student who identifies him-/herself only by the net-name, "LackLizard," over a suggestion that, in a utopian society, it might be necessary for countries to keep nuclear weapons to threaten "outside aggressors" [8/3] (note: the numbers in brackets denote "number of the thread"/"number of message in the thread" [see Table 1]): "How can we say we have a utopian society in which we are willing to employ a weapon that indiscriminately kills everyone—including noncombatants (especially children)—within its considerable range?" [8/4]. Mr. Winston (perhaps the most frequent poster to the BBS) also frequently uses such encouraging phrases as, "You have made a very important point..." [10/4]; "Great to see you contributing your insights" [1/5]; and "Does anyone have any ideas about how to deal with this issue?" [8/1] (Note: all quotations from the BBS are recorded exactly as they were originally written—no corrections as to grammar, syntax, or spelling have been made.)

At the same time, it is clear that a considerable amount of consumption is going on in the production subsystem. Demands on Mr. Winston and other teachers to keep discussions on track are satisfied at a cost to their energy; not only must they monitor the list with far greater care than the students (because the DORADO board is a BBS, not a mailing list, I am assuming that participants are not automatically notified of each post), but they must judge the appropriateness of posts and respond in ways that simultaneously incline students in the direction of desired pedagogical goals, while at the same time sufficiently acknowledging the poster's right to his/her own opinion (thereby encouraging free expression in the future).

The cost in terms of energy is frequently related to the fact that teachers must mediate between what they want to teach (academic knowledge) and what the students—based on their cultural learning—either can or want to grasp. This all-too-common pedagogical problem is exacerbated when students of different nationalities, not to mention different subcultures, are involved. DORADO BBS messages are replete with examples of a teacher advancing a provocative, challenging idea, only to have it misinterpreted (as with LackLizard's ill-considered remark about nuclear weapons, in response to Mr. Winston's previous question about the role of the military in a utopian society [7/1]). Even more draining of a teacher's energy, perhaps, are those instances in which a very interesting and provocative idea is posted and no one responds. In a dramatic example of this, another DORADO site organizer, Mr. Jaworski, referring to Delmore Schwartz's outstanding short story, "In Dreams Begin Responsibilities," raises these challenging questions: "Is it possible to dream without inviting responsibility? I mean, can one simply dream of a better society without taking action, or is the dream itself an action that invites new responsibilities? If so, at what point are we responsible for our dreams and their fulfillment?" [4/1] It is astounding that no one picks up this thread (not even other teachers), and the interesting question with respect to the activity system is, "Why?" I would argue that at least part of the reason lies in the inability of the instrument (the Internet) to convey subtlety, particularly when it has to do with an idea students might grasp only dimly, if at all. I would argue further that the likelihood is high that Mr. Jaworski's post might simply have gone over the heads of his young audience, which is another way of saying they had not been given, through cultural learning, sufficient conceptual tools to understand the point he was trying to make—or, perhaps more precisely, they may not have felt confident enough to take up the topic for further discussion on a public BBS. This appears to be confirmed by the fact that two of the DORADO threads—"Television and culture," and "Do we need the Internet?"—together account for an impressive thirty-five messages, more than half the total for 1998, even though both threads were judged to be of only moderate ("neither high nor low") relevance to the topic (see Table 1). In other words, while students were eager to offer their views on peripheral subjects such as television and the Internet (often straying far afield from the central topic, utopianism), they gave the far more pertinent topic, "dreams and responsibilities," a wide berth, thus placing further demand on DORADO teachers to keep the discussions on track.

In the consumption subtriangle (see Figure 2, subtriangle circumscribed by subject <-> object/outcome <-> community>), what is produced is the engagement of a community of various formal and informal educators in the service of the object, or the students' learning. What is consumed, on the other hand, is the advantage students might gain from utilizing subtle shades of cultural meaning, so that they could provide an acceptable display of knowledge appropriately expressed.

There are many ways this expenditure can occur, the most obvious being that, although much of the discussion is by residents of Estonia (together with a smattering of posters from other non-American countries), they are compelled to express themselves in English, so that finer renderings of meaning in their own cultures, best expressed through their native languages, are closed to them. This is evident as some of the

Estonian students struggle with English phrasing. Katia, for example, is apparently drawing an effortful conclusion about television and violence when she writes, "But I can tell you that my little brother is trying to make same kind of things like mouse and cat in cartoon 'Tom & Jerry'" [2/11]. Liliana, a 16-year-old girl, offers the following comment on "bad sides" to the Internet: "But if you buy yourself don't use those programs then they can't disturb you. What the other people do is their own business" [9/8]. A male student, Mikhail, responds to a post from Mr. Winston questioning whether democracy is best for a utopian society: "In my last letter I ment democracy in the most wide way, you went into details. Would you like to be under the dictatioral rulement? Maybe you have to choose the other country where do live? (don't get angry, I don't mean bad)" [10/3]. It seems clear that, in each of these instances, lack of facility with written English is preventing the Estonian students from making quite substantive points. If we can visualize their thought processes as vast reservoirs of material for analyzing and responding to tough questions about difficult topics such as utopianism, then surely any requirement that they force themselves to make their point in a language they do not understand very well must be seen as a consumption of the very material which would best permit them to frame an appropriate answer. Likewise, a less-than-skillful student response can partly consume the teacher's energy: not only does this make it more difficult for the teacher to determine if s/he has made a point successfully, but it requires the teacher to formulate another message or messages to correct the original, unclear expression.

Another form of consumption of energy evident in the consumption subsystem comes from the expenditures that happen when two people from different cultures obviously have different meanings of key terms. This places demands on teachers, students, and educators (both in and outside school settings), as the group struggles to discover meanings for important terms that will satisfy people of different cultures ("different" cultures conceived of here, not simply as different national cultures, but cultures which reflect other dichotomies as well—young people versus older, "professionals" in the realm of ideas versus "amateurs," heavy Internet users versus occasional or neophyte users, and so on).

A fascinating instance in which DORADO members differ in their use of a key term is to be found in a dispute over what "democracy" means. Valery, a male Estonian contributor (whether Valery is a student or teacher is unclear), begins the discussion by quoting an earlier e-mail message in which Mr. Winston argued that democracy might not be the best form of government in a utopian society: "We now live in a world where democracy seems to be almost taken for granted as the best form of government, yet few seem to feel we are heading for utopia. Is democracy so great after all?" [quoted in 10/1] In a reply to the post in which Valery quoted him, Mr. Winston ventures further: "What if... freedom allows people to pass laws that allow our rivers to be polluted or our forests to be cut or some children to starve all so that the majority can fulfill their desires? We have to ask if fulfilling our desires is the best way to govern" [10/2]. Mr. Winston further proposes that perhaps a utopian society might be better off ruled by a platonian philosopher-king, to which another male Estonian student, Mikhail, retorts: "I still don't believe that everybody will live better, if one man rules all the country. You said that when there is freedom people pollute nature and cut the forests. People pollute nature anyway" [10/3]. Mr. Winston (an experienced teacher who no doubt recognizes the need to keep a good debate going) continues with, "Is Estonia being ruled more wisely now that it is a democracy or is it just the freedom that came with it that is attractive? Is democracy required for freedom?" [10/4]. Next there is a response from 17-year-old Estonian Tanya, from a religious perspective: "God's Son told his people that He was their king. No one could bear that news and everybody today know what happened: they killed him in a pretty horrible way. It was 2000 years ago and yet I think people would act pretty much the same (except for the killing part)." [10/5] Mikhail ends the thread [10/6] with a fairly standard recitation of the virtues of democracy and freedom.

From this exchange, two conclusions can be drawn: (1) communication over the meaning and value of "democracy" is somewhat impeded, almost certainly because a teacher from a country where democracy is taken for granted is trying to communicate an abstract point about democracy to students of a country where democracy has only recently been instituted; and (2) considerable effort has to be exerted by both teacher and student to find a common ground over Mr. Winston's very controversial proposal. It is the energy-consuming work of remaining assertive in the role of teacher (the better to ensure the successful outcome, concept mastery), while taking account of differing views among the "educators" of the students (the community), that constitutes another way to account for consumption within the consumption subtriangle.

In the exchange subtriangle (see Figure 2, subtriangle circumscribed by subject <-> rules <-> community), production can be defined in terms of the evolution of the rules which have resulted in the development of DORADO BBS, together with the evolution of cultural learning mechanisms in the student's home culture(s). The product here can be conceived of as any agreed-upon guidelines provided to the student, either as part of the DORADO "experience," or as part of acculturation, which can be used to effect the final goal, concept mastery. Such guidelines have not only come about incrementally, but have resulted from extensive testing by trial-and-error (although of course much more has been involved in the development of the student's native cultures, than in the development of either the Internet in general, or the DORADO group in particular).

Consumption in the exchange subtriangle involves student and teacher drawing upon these guidelines for the purpose of having the student fashion an appropriate expression which demonstrates s/he has learned what the teacher intended. Neither the Internet culture, nor one's native culture, should be seen as static, reified entities, but rather as vast pools of cultural practices which simultaneously draw upon structures and in turn reinforce these structures through the repetition of social behavior (similar to Giddens' [1979] structuration theory).

In the exchange subtriangle, it is useful to view cultural resources as subject to appropriation by various members of the community of educators. Although we might take this immediately to mean, for example, that teachers rely primarily on the rules of the culture of pedagogy, whereas informal educators rely primarily on the rules of the student's native cultures, it is obvious that there can and will be cross-fertilization, with both formal and informal educators each depending on the cultural systems of the other to come up with ways to provide instruction to students. Thus, consumption occurs as educators appropriate, modify, negotiate, or reject cultural rules they deem necessary to use in the service of teaching. This process consumes an extraordinary amount of effort, particularly when there are significant differences between the cultures of formal and informal educators. Nevertheless, because cultures are active, dynamic, and living, they comprise "open systems" which benefit from the infusion of new ideas. These new inputs often provide the energy that the subsystem must consume in order to thrive.

As an example of teachers taking into account how cultures interpenetrate, each modifying the other, consider the following exchange. Mr. Jaworski, one of the site organizers, quotes Czechoslovakian president Vaclav Havel on the possibility of a "mission statement for humanity": "'It is not enough to invent new laws, institutions or machines,' Havel said, 'one must understand the true purpose of our existence and how we fit in.'" Jaworski continues: "In many ways this sounds like a request for humanity to write a philosophical 'mission statement.' I'm curious to hear what others think of this idea. Would it be helpful to create such a document? If so, how would a multi-cultural, multi-tribal group collaborate on it?" [1/1]. To this intriguing suggestion, Jennifer, an American student, replies: "Some people would get frustrated, some angry, and some would probably leave and go home. It would be a long, drawn out process that would move slowly because of the ideas that would be discussed. But, in the end, you would have a well-crafted, well-thought-out document that would provide a world view of humanity" [1/2]. Another teacher, Mr. Winston, is quick to reframe Jennifer's response: "I'm not so sure a mission statement would need to be so long and detailed, though. But then, maybe we could still use a revolutionary document as an example—The Declaration of Independence" [1/5]. Timothy, a student whose country of origin is unclear, questions Jennifer's predictions: "You mention in your essay that 'Some people would get frustrated, some angry, and some would probably leave and go home.' What would become of these unhappy persons? Less than a century after the Constitution was formed there was a war which I suspect had something to do with people who were angry and frustrated with it. Would the mission statement really be for Humanity if there were humans so displeased with it as to pack up and leave?" [1/3].

The various undercurrents in these e-mail interchanges can be profitably analyzed as problems of consumption in the exchange subtriangle. It is clear that the teachers (Mr. Jaworski and Mr. Winston) have very definite ideas about the "mission statement": Jaworski says it is an instantiation of Havel's thought (a questionable assertion, by the way); Winston thinks it could be a short document; Jaworski views potential collaborators as part of a "multi-tribal" resource pool; Winston thinks perhaps the American Declaration of Independence would serve as a good model; and so on. By their statements, these two teachers reveal the foundations of the cultural rules upon which their judgments are based. Jennifer, on the other hand, as an American student, comes to the discussion with some cultural rules of her own, revealed in such curious assertions as, "Some day, this world will either unite or destroy itself," offered without evidence in the manner one would expect from a young, inexperienced student. While Jennifer's rules for cultural interpretation incline her toward an optimistic view of conflict, Timothy's rules make him more pessimistic, so that he views anyone who would disengage from the writing of a mission statement as (at least potentially) one who could derail the whole process.

These divergences reveal that cultural rules are negotiated, through the expenditure (consumption) of energy. If we view rules as dynamic, then we have in the exchange subtriangle a three-way modification going on, in which subject, rules, and community interactively alter each other.

The final subtriangle, distribution (see Figure 2, subtriangle circumscribed by object/outcome <-> community <-> division of labor), brings us to the point at which what has been produced by the overall system, in toto, must be parceled out, according to the rules of the system, as espoused by the community. For the formal/informal teaching activity system, the distribution subsystem involves disseminating the object/outcome (learning) by members of the community (educators), through the efforts of various acculturators whose specializations lead to a division of labor. Production in this subsystem occurs as the community's standards (divided up according to the specialized acculturation tasks reflected on the division of labor node) are brought to bear on the object/outcome. What is produced is interpretation of the posted messages, often indicated (as in the “humanity's mission statement” example) by publicly displayed responses. Indeed, it is only in publicly posted responses that one can see any indication that cultural systems are incommensurable, and this incommensurability can be assumed to arise from conflict between the evident cultural underpinnings which give rise to the ideas expressed in posts and responses.

Consumption in the distribution subsystem comes from effort expended by those who perform specialized acculturation tasks (according to the division of labor) in trying to make sure the community (educators) perceive the posted message in (what each specialist takes to be) the proper framework. As was evident in the dispute over “humanity's mission statement,” any thread that stays “on topic” usually consists of a series of messages that advance analysis of the point under discussion—to compose a reply, one must begin by reframing the message one is replying to, generally in terms which make elements of the former message easier either to support or to refute. Consciously or unconsciously, the terms under which that reframing will take place are spelled out by the formal and informal educators (community) and therefore represent resources which must be consumed in order for the activity in the distribution subsystem to take place.

In summary, the production/consumption paradox for the overall formal/informal teaching activity system can be stated as follows: although the overall system is geared to the production of an acceptable outcome (concept mastery), the system consumes the following: (1) teacher energy (production subsystem); (2) advantages given students through native cultural elements (consumption subsystem); (3) effort in negotiating cultural resources, gained both in and out of the classroom (exchange subsystem); and (4) effort on the part of acculturation agents to reframe often conflicting messages in terms which are (to them) more appropriate (distribution subsystem). More briefly, the paradox is: while the system aims at producing what appears to be a unified, monologic message as proof of a pedagogically acceptable outcome, the various elements of consumption guarantee that the apparent unity in the outcome is an illusion. In other words, the message at the object node arrives to public view, not whole, but scarred through the course of the various battles it has fought to get there.

Step 3: Identify primary contradictions. The next step is to specify the contradictions on each of the nodes, thus further refining the ideas covered in the previous steps. As noted earlier, contradictions represent the manifestation of the paradox between production and consumption, and serve as key reference points for the activity theorist to devise a strategy to move the activity system toward new states of development. Primary contradictions are those within each constituent component of the activity. In Figure 3, based upon the analysis in step 2 (above), I have specified by the primary contradictions for the formal/informal teaching activity system.

As in all activity systems, the imbalance between production and consumption is what gives rise to a need state in the old activity system, making it necessary for a new activity system to come into existence. When we examine the formal/informal teaching system in terms of its primary contradictions, it becomes clear how difficult is the task of teachers who want to use the Internet to forge global communities. Beginning with the subject node, we find the teacher faced with a classic problem in pedagogy: how to get students to respond in an acceptable way without doing violence to their sense of personhood, tied so often to the various cultures into which they have been assimilated. At the instrument node, we find the Internet, as a form of communication linkage, conceived of in two ways. First, the Internet can be seen as a more-or-less objective purveyor of information, in which messages, denuded of many of the markers of social context, are presented in the cold, implacable light of basic text upon a screen. This conception of the Internet would seem to be best for posting pedagogically acceptable viewpoints. However, a second way of seeing the Internet is as a means to convey vast amounts of information reflecting one's cultural knowledge, not only directly (as when BBS participants make overt assertions about their native cultures) but indirectly (as when the juxtaposition of messages in a thread reveal underlying conflict over cultural meanings). At the subject node, one finds the primary contradiction over whether learning is to be regarded as having been achieved by the student formulating a pedagogically appropriate post, or by formulating a post which makes sense in terms of the various cultures of which s/he is simultaneously a member.

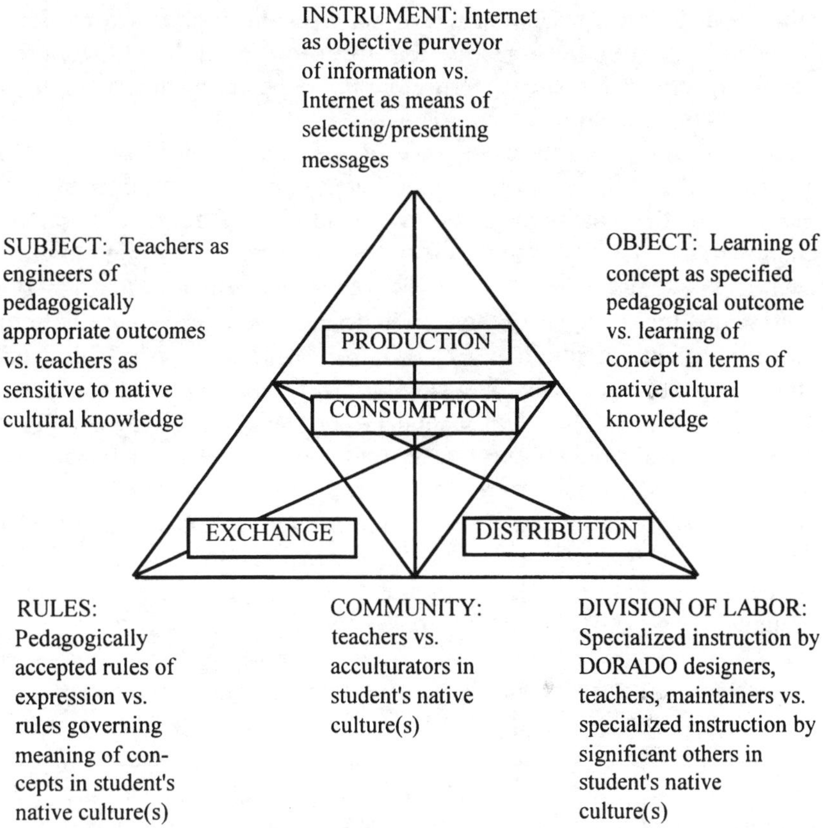

Figure 3
Primary Contradictions on Activity Triangle Representing Formal/Informal Teaching System, DORADO Group

At the rules node, pedagogically accepted rules for expressing ideas contradict native cultural rules governing how these same ideas are to be expressed. The community node encompasses those who provide the resources students will use to frame the expression of their ideas on the BBS; these individuals will (according to the way in which this activity system is conceived) be either, (1) the DORADO teachers (who directly regulate content through the authority of managing formal classroom procedures), or (2) the many people outside classroom settings who provide (and have in the past provided) instruction to the student in how to express ideas. Finally, the primary contradiction at the division of labor node reflects specialized instruction in the written expression of ideas performed by those responsible for teaching the DORADO group, versus specialized instruction performed by significant others in the student's native culture(s).

Step 4: Demonstrate emergence of secondary contradictions. With identification of the primary contradictions, we can now proceed to the real crux of the work which will eventually enable us to effect positive change in this activity system: identifying key secondary contradictions. Secondary contradictions occur between the primary contradictions represented by the nodes of the activity triangle; the need states that grow out of secondary contradictions give rise to springboards which can be used to move the activity system toward a more optimal state of development. Of the several secondary contradictions identifiable in the formal/informal teaching activity system, I will focus on two which are directly related to the role of the Internet in global society: (1) the contradiction between the instrument and subject nodes (delineated by the double arrow labeled "A," in Figure 4, below); and (2) the contradiction between the instrument and division of labor nodes (delineated by the double arrow labeled "B").

The secondary contradiction (arrow "A"), between instrument and subject nodes, relates primarily to limitations on the Internet as a means of expression. Although the DORADO teachers must use the Internet as a mediating instrument, it is clear from the primary contradictions at these two nodes that the Internet is not an optimal tool for teaching. The Internet cannot do much in the way of assisting the teacher to be sensitive to native cultural knowledge, when there must be so much selection regarding how that knowledge is expressed.

There are several ways in which the Internet's limitations make expression of teacher sentiments difficult. For one thing, BBS technology (at least, the technology available to the DORADO group) supports only written communication in one specific alphabet at a time. This means that even the teacher who wants to be sensitive to students' expression of ideas in terms faithful to their own culture(s) is constrained to offer instructions and feedback in a linear, relatively impersonal fashion expressed in English, the language of the dominant culture. No matter how carefully the teacher phrases messages, there is always the potential for misunderstanding to occur, not simply over grammatical rules, but over the considerably more significant issue of idiom. In face-to-face communication with a student, at least the teacher can deduce the student's lack of understanding through such nonverbal displays as the puzzled frown, the nervously tapping foot, or the crossed arms; with Internet communication, no such feedback is possible. Instead, the whole of feedback is expressed through written English messages.

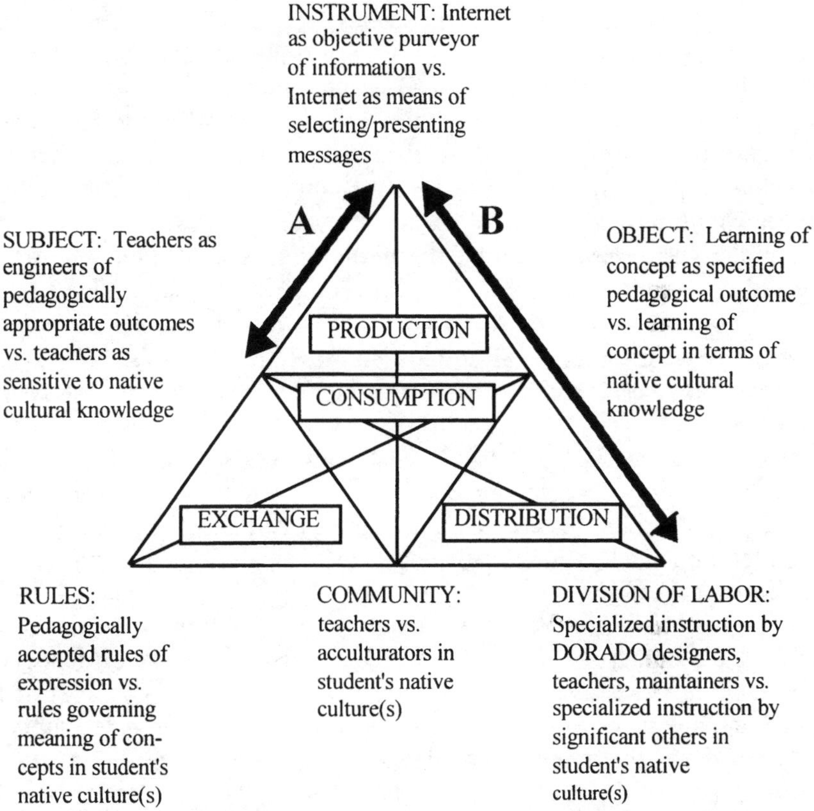

Figure 4
Primary and Secondary Contradictions on Activity Triangle Representing Formal/Informal Teaching System, DORADO Group

It can of course be argued that written English expression may in fact be the "pedagogically appropriate" outcome referred to in the subject node's primary contradictions (see Figure 3). In other words, it is not too much to assume that, in DORADO's group, one big reason for non-American students to participate is precisely because the communication is in English. Such an arrangement permits students to practice writing English by corresponding not only with native English speakers, but with students from other countries who are also learning the language. This would appear to satisfy everyone's needs and to be an acceptable way of communicating.

However, consider how much constraint is placed on the teacher who must rely exclusively on English to sense the meanings underlying student messages. In the DORADO group, one notices a very different tone of expression between American and Estonian students, and (perhaps unconsciously) a difference in the tone in how American teachers respond to students of the two countries. Notice the playful, rather charming air of Mr. Winston's reply to Jennifer's post about the Constitution: "Great to see you contributing your insights again this year. I like your analogy of the Constitution... Of course, one of the problems there would be to find a modern day Thomas Jefferson to eventually go off and put all the ideas into eloquent words. Would you like to volunteer?" [1/5]. Jennifer is equally charming in her reply: "Well, Mr. Winston, if you get all of the ideas together, I will put them into 'eloquent words' :-)" [1/4]. Although Winston is always polite when responding to Estonian students, he is seldom that friendly, and usually a little more distant. One reason may be that, when he addresses Jennifer, he is addressing a native speaker of English with whom he can communicate easily.

Another manifestation of the secondary contradiction between subject and instrument arises from the fact that most channels of feedback (feedback is what gives the communicator information to appropriately formulate messages to achieve the object/outcome) are closed to subject. A particularly thorny problem of CMC has always been that, once a message is sent, the sender (unless s/he speaks to the receiver personally, as by telephone or face-to-face) has only one way of knowing whether it has been received: the receiver must formulate a reply, acknowledging the sender's message, must post that reply, and have that reply read by the original sender.

Nevertheless, it is evident that this configuration of messages—responses makes it difficult, if not impossible, for DORADO teachers to achieve the object/outcome of the overall activity system. Let us suppose that a teacher were to post a message to the BBS, and no one was to reply. What does such a "non-response" mean? Does it mean no one read the message? That people read it, but did not consider themselves competent to reply? That they agreed and thought no reply was necessary? Once the message is sent through the BBS instrument, it is impossible to determine which (if any) of the above possibilities could explain the lack of response. This makes it particularly difficult for the teacher (subject) to contemplate the next move in trying to achieve the object/outcome. Indeed, if the teacher attributes the lack of response to the wrong causes (for example, insolence or indifference), and publicly states so, that might damage the entire sense of community on the BBS.

Finally, the secondary instrument/subject contradiction manifests itself in culturally-prescribed rules of selection and presentation. Despite the fact that the Internet is supposed to be a more-or-less objective purveyor of information, it remains that the information provided through the Internet must be selected by the subject according to culture-specific rules. In other words, nothing can be presented through any Internet venue without first having achieved some cognitive priority in the mind of the person formulating the message. The sender's cognitive processing hierarchies will prioritize certain message formulations, dismiss others as irrelevant, and will not even conceive of still others. Moreover, the decision to present a certain message in a certain way will be constrained by many factors in the experience of the message composer, and the most influential of these will be rules assimilated by the composer, through the various cultures s/he has experienced.

This is of course a problem of which every teacher is aware, and one that all teachers must struggle with as they search for the most optimal message to allow them to convey knowledge to the student. However, when the instrument mediating teaching is the Internet, complicating factors arise. For example, it is well known that certain cultures prefer circumlocution, while others, uncomfortable with "beating around the bush," need messages to be expressed directly and without delay. American culture favors the latter approach, which seems to be ideally suited to Internet communication, particularly on a BBS: messages are characterized as occurring in "threads" (i.e., argument—response—argument), not by keyword or other classification scheme; messages are highly specific as to time and date (consonant with the M-time [Hall, 1983] orientation of American culture); BBS communication fits very well with the rules for conversational turn-taking favored by Americans, with posters having to wait for other messages to be posted before they can be replied to; and so on. The point is that these restrictions do not have to await interaction to become problems; rather, they are factors taken into account by the composer of messages prior to the formulation of the message itself (Bakhtin, 1986). Indeed, it is impossible to conceive of a teacher formulating a coherent message without being aware of the limitations of the Internet to convey that message in such a way as to effect the object/outcome. The problem for the teacher is that—in the absence of analytical mechanisms such as activity theory—the fact that s/he is making culture-based assumptions about communication might pass unnoticed, leading to the mistaken conclusion that something in the content of the message, or about the subject to which the message refers, is the source of any misunderstanding. Without an awareness that it is perhaps the nature of the Internet itself that may be impeding communication, the teacher might forever search for a false cause, unaware that the real cause of misunderstanding may lie in the mediating instrument.

These are thorny difficulties, to be sure. However, the value of activity theory is that it permits the analyst to probe even the most intractable of system contradictions, and from that vantage point, to propose some possible solutions. In the concluding section of this chapter, I will suggest some things the DORADO organizers might try; at this point, though, let us move on to discuss the secondary contradiction between instrument and division of labor (arrow "B" in Figure 4).

The instrument/division of labor secondary contradiction concerns the question of identity, and more precisely, what it is that comprises the identity of the person who can only be seen through the words they type on a computer screen. Division of labor, as we have conceived of it in this activity system, refers to specialized tasks whose purpose is to teach the student to express him- or herself in English by means of the Internet (instrument). The phrase, "express oneself," refers to the process of making manifest in writing, and thence through Internet posting, certain facets of oneself that one wishes to be known. An often-repeated criticism associated with Internet "identity" is that there is no way one can be sure what the poster says about him- or herself is true. As recent arrests in cases of child sexual solicitation over the Internet have dramatically proven, one can assume virtually any identity in cyberspace—police officers have caught pederasts simply by pretending to "be," online, children under the age of consent.

Why, then, could we not suspect that, on the DORADO board, we are getting a false picture of the object/outcome? When students write of their experiences, what keeps them from making assertions that are not true, thereby giving a false impression of the degree to which they have achieved concept mastery, the goal of the overall activity system? It is because the socialization elements at the division of labor node prevent that from happening. On the one hand, teachers, through collaboration among themselves—both (one presumes) through private e-mail and through the organizers' BBS—formulate educational strategies, evaluate events, discuss options, and propose new ideas. Through such activities, teachers both monitor student performance, and serve as socialization agents, and the student who expresses concept mastery must take account the constellation of DORADO personnel, knowing that his or her educational success depends on performing in a way the DORADO teachers consider appropriate.

On the other hand, however, student exhibitions of concept mastery are equally constrained through instruction provided by significant others in the student's native culture(s). One type of such informal instruction are some messages to the DORADO board which are replete with indicators that the opinion of the student poster is being constrained by the opinions of other students, known personally to the poster. As an example, consider the back-and-forth involved in the following exchange about popular music by several students from Estonia. The thread is initiated by Anatoly, a 17-year-old male Estonian student (unusual in itself, since most threads are initiated by teachers): "Strange music is music that you can't hear on TV or Radio during the day and this music is not for everybody. It's published in small number of copies and you can buy it only in the specials stores" [6/1]. To this, Ludmilla, a 17-year-old Estonian girl, replies, "I listen every kind of music, starting with classical and finishing with that 'strange' music. I like them all. I have friends who does not want to listen that kind of music. But I do respect them and doesn't judge them because of it. I also have friends, and they are good ones, who make 'Underground' party's, make this 'strange' music and I'm enjoying it with them." [6/2] Estonian student Alexia rejoins, "I think you are absolutely right about this. It is good to be original and different from other people in some things" [6/3]. The thread continues in much the same vein for the remaining six messages, with students from Estonia, as well as two students from Mexico, reinforcing each other's views regarding the cultural influence of music.

Less evident, but nevertheless influential, are the indicators in messages of the effects of family, folk symbols, political leaders, and other important conveyors of cultural education. For example, the last thread (a single message, really) of the 1998 DORADO messages contains this moving passage from a 17-year-old Russian girl, Natalya, concerning two monuments in Estonia to victims of shipwrecks (the "Russalka" and the "Estonia"): "Many ships had found their death in the sea during these 100 years, but these two are living in our memory and have the monuments... Who knows, maybe the deaths of Russalka and Estonia were the 'special' events or signs for peoples? Ship—this is a symbol of dream, aim and soul. The ship's ruin is a loss of direction... Do we really have to create the third monument?" [12/1]. Clearly, this young woman has been profoundly affected by a variety of complex symbols in her culture, and has made the leap from these symbols to genuine insight into the movement of history in the last years of the twentieth century. Such a message would not be possible were it not for the complex interweaving of countless elements of cultural education provided her through the manifold cultures that she has experienced.

Thus, it is clear that no message makes it to the point of representing the object/outcome node without having been mediated through many cultural filters. However, the problem is that, although the presented message appears to be a monologic text, expressing what teachers must interpret (at least publicly) as an outcome of student learning, in fact the text is shot through with hidden conflicts, unvoiced assumptions, and concealed influences. Yet when the teacher responds to posts indicating students' concept mastery, the venue through which the teacher expresses him- or herself (i.e., the Internet [instrument]) makes it appear that s/he regards the student posts as monologic, sequential, and addressed primarily to a single topic, when in fact student posts may be (and often are) dialogic, out of sequence, and addressing several topics at once.

An example of this process occurs in the thread, "Television and culture," which received more posts (20) than any other thread on the 1998 DORADO BBS. While it was apparently difficult to get members to comment on some of the more abstruse issues raised in various threads, many people wanted to contribute their opinions about television, with Estonian and Mexican students in particular raising a large number of interesting points about how television, particularly imported American television, affects their lives.

Many of the posts on the thread are densely packed with ideas that deserve to be appreciated for their subtlety and depth. For example, consider this rather ineptly expressed, yet profound, observation from an anonymous Estonian student: "Does TV effect on our society? I think that it does. But is it positive or negative it dipends on what kind of programmes do we wach. If we wach soaps, the effect has to be negative. You ask why? That's because if we wach THINGS like that we might thik, that life is simple as in soaps or even we do not think at all. And why do not we think at all? Because all our possible thoughts are allready on the screen" [2/5]. Another Estonian student, Lena, links television to other forms of media: "Most people that I know watch TV and read magazines every day and we are constantly discussing about the broblems we've seen or read. It seems to me that we almost do everything as we have seen or read it to do" [2/10].

Nevertheless, when Mr. Winston comes to the point of summarizing the messages on the thread, we get the following: "Hi everyone (especially all of you Estonian students who have contributed so many thoughtful messages to this thread): There are too many ideas that have been expressed here to comment on all of them, but one seems to be pretty well accepted by everyone who has contributed here: TV is not good for young children. My question is, why do their parents let them watch it if it is not good for them? Are you saying that you recognize the bad influence of TV and parents don't? Or do they understand also but just can't do anything about it? In either case, I wonder what effect this has on the family? I wonder if it diminishes the authority of the parents, since it seems to teach children behavior parents (we would hope) are against?" Mr. Winston's summary encapsulates dozens of issues raised in seventeen different posts (excluding the three Winston wrote for this thread himself) into what he takes to be a key issue ("television is not good for children"). Although his summary is fairly complete, it clearly fails to acknowledge the great majority of ideas expressed in the posts (a fact Winston himself acknowledges).

A good part of the reason for the premature closure of the students' diversity of opinion lies in the nature of the Internet itself. Because posters must write messages that appear in sequence, they tend to limit what they respond to. Not only is written communication of any kind more likely to "freeze" discourse process (committing a thought to writing places it on display for public inspection and makes it impossible to change), but the sheer volume of Internet messages often makes it impossible to respond to any but the most limited aspects of a posted message. Mr. Winston makes a valiant effort, but loses a considerable amount of detail in the process. Were Mr. Winston communicating with the students as a group, face-to-face, there is no doubt he could use his teaching skills to sense and respond to multiple channels of verbal and nonverbal feedback; coordinate several different topic "threads" simultaneously; use responses from one student to stimulate responses from others; and try to make sure everyone participates. Communicating with students over the Internet, however, he can respond to only one feedback channel; must usually limit himself to only one, or, at best, a few, topics; speculate about whether students are engaged by the discussion or not; and leave it up to students to decide whether to participate.

It seems appropriate to characterize these difficulties as a conflict between the Internet as a venue for communication (instrument node) and task specialization relating to training students to express themselves (division of labor node). On the one side, we have the BBS, an extraordinarily limited way of expressing oneself, while on the other we have the complex interweaving of a multitude of ideas taught students about how to express themselves in discursive argument. The mismatch between the two nodes is one that will point the way to potential points of access as we attempt to move the formal/informal teaching activity system to a new level of development.

SUMMARY—SUGGESTIONS FOR DORADO GROUP

To conclude the analysis of this DORADO activity system, we turn now to the question of what can be done with the conclusions that have been drawn. The real goals of activity theoretical analysis are, first, to rigorously analyze the target system, and then, second, to apply the insights gained to the system, as it evolves, so as to observe what happens to it, both in terms of its own evolution (here we would look for "tertiary contradictions," between the "old" system and its "newer," more advanced form) and in terms of how it interacts with other systems (and here we would look for "quaternary contradictions," between the target system and other neighboring activity systems related to it). For the next stages of analyzing DORADO's formal/informal teaching activity system, I would suggest attention to the following three issues.

First, site organizers should make participants more aware that the messages they post are dialogic in nature. Organizers should emphasize that even the simplest message contains many indicators of the richness of social life that has given rise to its expression in writing. Such richness should be celebrated and those who post should be encouraged to make their messages more, not less, complex. By following this suggestion, organizers might more successfully deal with the secondary contradiction between the instrument and subject nodes (arrow "A," Figure 4), as well as the secondary contradiction between the instrument and division of labor nodes (arrow "B," Figure 4).

Second, site organizers should extensively brief participants on the limitations of the Internet as a means of conveying information. Posters, whether students or teachers, should be made repeatedly and clearly aware of how Internet communication limits and constrains discussion. Indeed, such revisions could serve as a palliative to widely publicized, but mistaken, portrayals of CMC as an ideal form of communication for the exchange of ideas. Attending to this suggestion, organizers might more successfully deal with the secondary contradiction between the instrument and subject nodes (arrow "A," Figure 4).

Third, site organizers should make every effort to elicit background information from site participants relevant to the experiences that brought them to the Internet. Students and teachers alike should write detailed biographies about themselves, their interests, and who and what has influenced them the most; these biographies should be linked to each and every message by hypertext (perhaps as part of an automatically appended signature file). This would enable those who read the posted messages to "fill in" the poster's background, thereby rendering a more complete portrait of the message composer. To ensure greater attention to the biographies, all posters should be encouraged to make reference to background when replying to a post (if necessary, teachers should include in grading and other evaluations how well respondents successfully pay attention and incorporate information from the biographies in their replies). Attention to this suggestion might allow organizers more successfully to deal with the secondary contradiction between the instrument and division of labor nodes (arrow "B," Figure 4).

If we were to move the activity system forward, we would take these three suggestions, incorporate them into the DORADO procedures, and observe the effects. As we gained more data about how these revisions of procedure worked to modify the activity system, we would compare the findings about later evolutionary states in the system to earlier ones, and thereby gather information to find out whether or not our procedural revisions had worked.

IMPLICATIONS FOR GLOBAL SOCIETY

The findings concerning the formal/informal teaching system of the DORADO group have a number of important implications for the study of whether it is possible to achieve global communities by means of the Internet. By attending to these matters, we will make more likely the outcome of global community, with the Internet as an important factor in achieving that community.

First, these findings emphasize how important it is that new, more interactive technologies be devised for the exchange of information. It is clear that much of the difficulty of Internet communication can be laid at the doorstep of sequential, written postings. Perhaps, at some future time, the technology will exist to convey messages in more than one language, or even to use multimedia (e.g., video and sound projections) to let posters and responders see and hear one another. Global community depends on our willingness to encourage innovation in the technical sphere and not to be afraid of technological advancement, so long as it is achieved with assurances that it will be used in the service of bettering human life.

At the same time, technology cannot be entirely relied on to solve the underlying problems of global society we saw exemplified in the DORADO system. Technological innovations are usually expensive, and many of world's people either cannot afford "state-of-the-art" technology, or else the communication infrastructure in the places where they live make it impossible to utilize such technology even if it were available. Therefore, we must not lose sight of the importance of "lower-tech" solutions which would allow us to "color in" the outlines of the people we see only in posts: mailed letters and pictures; writings in books about other people and cultures; videotapes, both of locales and of individual people who inhabit them; and so on. The lure of Internet communication lies in its comparative ease; but therein also lies its greatest danger. Often, more effortful activities made to get in touch with the cultural "other" reap greater rewards. Global community depends on using the Internet as one tool to achieve mutual understanding, but never as the primary or only tool.

Finally, along with using the Internet as a means of forging trans-cultural communities, we need to make a concerted effort to advance our understanding of cross-cultural interactions. This can be achieved by redoubling efforts to include cross-cultural classes in primary, secondary, and post-secondary learning institutions; cross-cultural training in business and industry; and broad-based, government-supported, campaigns of public education (principally through media such as television and radio) to increase public awareness of the problems and potentialities of global communities. It seems that most advertisers, and non-commercial organizations as well, include in media campaigns directions on how to get to their Internet websites or how to communicate with them via e-mail. It is time that these same procedures be instituted in the service of achieving greater intercultural understanding and thereby enhancing the likelihood of achieving global community

REFERENCES

Bakhtin, M. M. (1986). The problem of speech genres. In C. Emerson & M. Holquist (Eds.), *Speech genres and other late essays* (pp. 60–102). Austin: University of Texas Press.

Bakhurst, D. (1988). Activity, consciousness and communication. *The Quarterly Newsletter of the Laboratory of Comparative Human Cognition, 10*, 31–39.

Bruner, J. (1985). *Actual minds, possible worlds*. Cambridge, MA: Harvard University Press.

Cole, M. (1985). The zone of proximal development: Where culture and communication and cognition create each other. In J. V. Wertsch (Ed.), *Culture, communication, and cognition* (pp. 146–161). Cambridge: Cambridge University Press.

Engestrom, Y. (1987). *Learning by expanding: An activity-theoretical approach to developmental research.* Helsinki, Finland: Orienta-Konsultit Oy.

——. (1990). *Learning, working and imagining: Twelve studies in activity theory.* Helsinki: Orienta-Konsultit.

Engestrom, Y., & Middleton, D. (1996). *Cognition and communication at work.* Cambridge: Cambridge University Press.

Giddens, A. (1979). *Central problems in social theory: Action, structure and contradiction in social analysis.* Berkeley, CA: University of California Press.

Glaser, B., & Strauss, A. (1967). *The discovery of grounded theory.* Chicago: Aldine.

Hall, E. T. (1983). *Dance of life.* New York: Doubleday.

Holt, G. R., & Morris, A. W. (1993). *Activity theory and the analysis of organizations. Human Organization, 52*, 97–109.

Leontyev, A. (1978). *Problems of the development of the mind.* Moscow: Progress Publishers.

Luria, A. (1976). *Cognitive development: Its cultural and social foundations.* Cambridge, MA: Harvard University Press.

Marx, K. (1973). *Grundrisse: Foundations of the critique of political economy.* Harmondsworth, England: Penguin Books.

Vygotsky, L. S. (1978). *Mind in society: The development of higher psychological processes*, M. Cole, V. John-Steiner, S. Scribner, & E. Soubermann (Eds.). Cambridge, MA: Harvard University Press.

CHAPTER EIGHT
Global Communication via Internet: An Educational Application

Guo-Ming Chen
University of Rhode Island

INTRODUCTION

The world is shrinking. People of different cultural backgrounds are more interdependent than ever. The 21st century will confront the ever-shifting social, cultural, and technological challenges. Rapid development in every aspect of the 21st century will demand us to see things through the others' eyes and to develop a new way of living together. Five trends have combined to promote a more interdependent future that shapes our differences into a set of shared concerns and a common agenda. These trends will transform the 21st century into the age of the global village in which people must develop a global mindset in order to live meaningfully and productively. They include: (1) technology development, (2) globalization of the economy, (3) widespread population migrations, (4) development of multiculturalism, and (5) the demise of the nation state (Chen & Starosta, 1996, 1998). These dynamics argue strongly for the development of more proficient intercultural communicators in this globalizing society.

The coming of new transportation and information technologies connects all nations in ways that were possible only in the imagination before this century. Communication technologies including the Internet computer network, facsimiles, the cellular telephone, interactive cable TV systems, and the information superhighway permit us instantaneous oral and written interchange at any hour to most locations in our country and the world. Porter and Samovar (1994) indicated that the improvement of information technology has greatly reshaped intercultural communication. It has created common meanings and a reliance on persons we may or may not meet face-to-face at some future date in our lives. Government is no longer the only source to disseminate information across cultural boundaries; indeed, common people daily talk and type their way into a web of mediated intercultural interactions. The immediacy of our new technology involves us with persons of different regions and ethnicity within our own nation and the world, and builds in us a new sense of national and global commonality. We find ourselves moving from unconcern, to curiosity, to an active need to improve our understanding of persons and groups from outside our immediate circle.

The progress of communication and transportation technology has made markets more accessible and the world of business more globally interreliant. It requires nations to decide how to remain competitive in the presence of new trade communities, and to try to find ways to promote products and services in other places where they have not historically existed. Such interdependence among national economies hinges on effective global communication, and calls for ever more skillful interaction across linguistic and national boundaries. Developing greater cultural and ethnic understanding becomes necessary both to carry out world business and to preserve threatened cultural identity.

Added to the cultural interconnectedness that accompanies technological development comes the influence of cultural migration between nations. Conditions at home and abroad push or pull persons to leave their country to find peace, employment, education, or a new start. The trend leads to a multi-ethnic composition of a society in which the contact among co-cultures becomes inevitable. Children in multicultural classrooms and workers in multinational corporations look for ways to learn and work efficiently in settings that are no longer defined exclusively by mainstream norms and rules. The quest for more productive interaction in international and domestic settings calls for further development of intercultural or global communication as a field of study which investigates the dynamics of interaction among persons of differing ethnic or national origin.

The development of multiculturalism has changed demographics of modern societies that affects every aspect of life (Adler, 1997). These demographic changes will produce classrooms and work places that are defined by no predominant ethnic culture or gender. The tributaries of different ethnicities, nations, genders, ages, tribes, and languages will flow into the mainstream of the classroom and workplace. Cultural diversity or multiculturalism will become the norm, not the exception. Global communication scholars will be needed to smooth the transition to bicultural, bi-dialectal classrooms, to multinational boardrooms, to multiethnic neighborhoods, and to gender and ethnic sensitivity on the part of professionals and service providers.

While new immigrants are arriving and co-cultures are making headway in achieving fuller participation, our very idea of our own identity will be sure to change. Increasingly, the de-emphasis of the nation state pulls us into regional alliances, such as NATO or NAFTA, that are larger than the nation. In addition, we see the re-assertion of ethnic and gender differences within the nation. To be able to negotiate meanings and priorities of diverse identities becomes a prerequisite of being competent in modern society (Collier & Thomas, 1988).

These trends have combined to provide a foundation for the indispensability of communication competence in the coming global society in which we are required to demonstrate "tolerance for differences and mutual respect among cultures as a mark of enlightened national and global citizenship" in individual, social, business, educational, and political institution levels (Belay, 1993). Thus, the need for intercultural communication enhancement, especially intercultural sensitivity, across disciplines becomes critical for success in future world.

These trends as well bring about a great challenge to American higher education: the challenge of globalization. College and universities, as the higher educational institutes, must provide an environment in which students can learn the nature of global society and learn skills for effectively communicating with people of diverse cultures. The key to the success in facing this challenge academically is to internationalize the curriculum by helping students (1) understand the goal of globalization in order to make their choices, (2) understand sensitizing cultural concepts that will assist them in their interaction with people from other cultures, (3) change aspects of their performance such as cultural self-perception and emotional and cognitive acquisitions in order to reach a higher level of empathy, (4) govern their performance and emotions in working and in living with people from other cultures by increasing their adaptability, and (5) adopt a changed way of perceiving and behaving so that they can improve their social performance in other cultures (Stewart, 1979). All these can only be reached through the enhancement of global communication by sensitizing students to understand, acknowledge, and respect cultural differences of people from diverse backgrounds. With the development of communication technology, the global Internet exchange provides us a great opportunity to help our students better equip themselves with knowledge and skills for survival in the 21st century.

INTERNATIONALIZING CURRICULUM VIA E-MAIL DEBATE

Electronic mail (e-mail) is one of the most common forms of computer-mediated communication (CMC) system today (December, 1996; Harasim, 1993; Metes, Gundry, & Bradish, 1998). Through a telephone line, computer, and satellite, e-mail messages can reach every corner of the world within several seconds. The e-mail system allows people to communicate in a one-to-one, one-to-many, and many-to-many formats. The system not only provides a powerful and non-traditional tool for students to learn through dialogue and collaboration, but also encourages "students to resist, dissent, and explore the role that controversy and intellectual divergence play in learning and thinking" (Cooper & Selfe, 1990, p. 849).

The e-mail system has been extensively applied at educational institutions through international networks (Internet) all over the world (Ma, 1994, 1998). Dern (1992) reported that there were over three million users with 500,000 computers in 33 countries using e-mail linked through Internet. According to Nua Internet Surveys (1998), this number has increased to 147 million users. Among them, about 87 million users coming from North America with about 53 million adults in the United States (Mediamark Research, 1998). Using the Internet e-mail system helps people overcome the constraints of time and space on geographically dispersed institutions (Garnsey & Garton, 1992).

Academically, the e-mail system connects campuses in different nations and provides students with an opportunity to communicate with their culturally dissimilar counterparts. According Ma (1993, 1994), exposing students to international e-mail communication leads to three effects: (1) Participants become better-informed about each other's culture, (2) participants disclose more than they do in face-to-face situations, and (3) e-mail communication is perceived to be more informational. The three effects indicate that students who participate in the international e-mail communication tend to be more open in sharing information which in turn provides them with subject and cultural knowledge. These effects were supported by Cohen and Miyake's (1986) study which shows joint participation in e-mail interaction across cultures can encourage multilingualism and awareness of other cultures. The educational potential of intercultural e-mail communication is enormous (Chen, 1994).

Thompson (1987), Chen and Kim (1993, and Shamoon (1998) pointed out that the e-mail network could help students work collaboratively, solve problems, and experience writing as communication in real situation. Hawisher and Selfe (1991) indicated that by thinking critically and carefully about the technology, instructors can successfully use the e-mail system to improve modern education, especially student's writing ability. Chen and Wood (1994) further suggested that using e-mail as a learning tool can help students learn the subject matter, improve writing skills, cultivate critical thinking, and reduce computer anxiety.

Based on this literature review, a highly structured project was launched using e-mail as a debate tool to internationalize college curriculum, with the rationale that with the globalization of the economy preparation for careers in business was becoming increasingly complex. Technical competence for a self-contained American market is no longer adequate. Students are required to have a broader, more liberal education, stronger writing and speaking skills, and knowledge and skills sufficient for effective interaction with their counterparts abroad. The trend demands that educators must explore more creative ways to challenge the traditions of compartmentalized and monocultural learning in order to accelerate the internationalization of the curriculum affecting all students. The educators must find ways to help students think critically, understand the world beyond borders, and communicate effectively across cultures.

The project was proposed to internationalize (1) each of the key undergraduate courses encountered by most business students, and (2) the manner in which each student learns through the use of an e-mail system linking every major research and teaching institution worldwide.

Ten key business courses, at a mid-size public university in the New England area, encountered by most undergraduates either majoring or minoring in business programs, were selected to be the experimenting courses for this e-mail debate project. Teams of course instructors in the experimenting courses were required to cooperatively revise their course to appreciate each other's cultural differences. Subsequently, students were expected to communicate with each other throughout their student years by means of e-mail debates on pertinent professional issues affecting each society.

Two long-term objectives of this project were (1) to better prepare students to function effectively as citizens of the global society and enjoy expanded career opportunities in the global workplace, and (2) to help students develop a network of international contacts. From these objectives, four specific short-term objectives were derived. They were to help students (1) acquire intercultural sensitivity by developing a working knowledge of international dimensions of business, and an ability to deal with various types of situations they will encounter in doing business with people from other cultures; (2) improve writing ability by reducing writing apprehension and developing an effective writing style; (3) increase computer competency by developing ability in using international computer networks; and (4) improve critical thinking by developing organizational and reasoning skills. In order to examine the impact of the e-mail debate on global communication, this chapter reports only the results of the project regarding the acquiring of intercultural sensitivity.

METHOD

Participants

With their counterparts in Denmark, France, Germany, Hong Kong, and Turkey, 430 American students in the college of business participated in this global e-mail debate project. The American students were asked to take the pre- and post-test on the measurement of intercultural sensitivity, both in Likert-scale test and open-ended questions.

Procedures

This international e-mail debate project took the whole semester to implement. A total of six semesters were used to complete the experiments. Prior to the project, all students in the class were required to learn how to send and receive a message in the e-mail system. On the first week of the semester students were organized into groups of three and a debate topic was assigned to each pair of teams—affirmative and negative terms. The class instructor served as the debate evaluator and judge. Examples of debate resolutions include:

(1) Japanese are soon going to achieve supremacy in computer technology that the U.S. currently enjoys.

(2) Individual rights and privacy are likely to be invaded in a society that is driven by information technology.

(3) The globalization process is accelerated by developments in information technology.

(4) The nature of marketing research needs to change little across Western cultures in order to be successful.

(5) Marketing research is more important to companies that compete in international markets.

The debate format was designed by a debate coach at the Department of Communication Studies. The debate was comprised of four arguments. The constructive argument was the first submission in the debate. In the constructive argument students advanced all the arguments that they wanted to make complete with reasoning and evidence. The refutation argument was the second submission in the debate that refers to clashing directly with an opponent's arguments. The rebuttal argument was the third submission in the debate in which group members defended and extended their constructive arguments, in light of the refutation made against their case. No new arguments were allowed in this stage. The executive summary was the last submission in the debate in which group members organized all the major arguments in the previous three submission files.

A two-week period of time was given for each of the first three arguments to prepare, organize, and send the file to their opponents. A week was given to the executive summary. The length of each of the first three arguments was 3,000 words, and 1,000 words for the executive summary. Each group was required to submit their arguments via e-mail to the opposing group by the due date.

During the first week of the semester, the instructor in each debate class explained the purpose of the debate project and distributed questionnaires to students. The data collected at this stage were counted as the pre-test results. During the last week of the semester, the instructor again asked students to fill out the questionnaires. The data collected at this stage were counted as post-test results.

Measurement

In order to answer the research question about intercultural sensitivity, students were asked to complete Chen's (1993) Intercultural Sensitivity Scale at the pre-test stage. The alpha coefficient for Intercultural Sensitivity Scale is .77 in this study. In the post-test stage, in addition to the instrument used in the pre-test stage, students were asked to answer six questions on a 5-point scale regarding satisfaction, critical thinking skills, computer skills, writing skills, and intercultural sensitivity. An example of the questions is "Did participating in the e-mail debate make you feel more sensitive to the other person's point view?" Moreover, four free-response questions were also used to solicit information about students' feelings on the e-mail debate project. Two of the questions relating to intercultural communication are "What aspect(s) of e-mail across borders did you like the most? Why?" and "Do you think this technique will help increase intercultural sensitivity among students? Please explain."

RESULTS

The test scores were computed to examine differences between pretest and posttest in terms of intercultural sensitivity. The results do not show significant differences in score. Nevertheless, most of the responses to the question on whether they thought the e-mail debate increased their cultural sensitivity were positive.

After content analyzed the four free-response questions, the participants' answers were classified into different categories, and representative answers were included to illustrate the categories. Answers for the two intercultural communication-related questions are reported here.

For the first question "Which aspects of the e-mail debate did I like the most? Why?" answers were categorized into three dimensions: communication, culture, and debate format. The following are sample answers:

1. On communication:

(1) Communicating with the students who live very far away makes me feel like I now have friends all over the world.

2. On culture:

(1) Seeing how individuals from a different culture and country view things.

(2) Learning about how people from different cultures have different values and ideas.

3. On the debate format:

(1) The topic of the debate which allowed for friendly competition.

(2) The group work involved in the debate. Learning the mechanics of using e-mail.

Sample answers for the question "In what way can the e-mail debate promote intercultural sensitivity?" include:

(1) Should allow students from different cultures to communicate on a regular basis over a longer period of time. One semester is too short.

(2) The debate on cultural or social issues with an international focus will enhance students' understanding of the importance of these issues to citizens of other nations. For example, debate on topics such as inter-racial marriage or immigration will enhance intercultural awareness and sensitivity.

(3) By providing opportunities for students to learn about the culture of their counterparts before the debate. For example, reading assignments or videotape will be helpful.

(4) Expand the "electronic handshake" to promote additional, informal communication between students.

(5) Form mixed teams for debate. That is, one can form a debate team by drawing students from several countries. The debate between such teams will certainly be more effective in promoting intercultural sensitivity than otherwise.

DISCUSSION

Through a highly structured format of computer interaction, part of this project was to examine the impact of global e-mail debate on intercultural communication. Both Likert scales and open-ended questions were used in this study. Although the attempt for measuring intercultural sensitivity by using Likert scales before and after the project proved much more difficult than we expected, the positive feedback based on the open-ended questions reveals that the e-mail debate showed a positive impact on improving intercultural sensitivity.

In addition to that, a 12-week project may be not long enough to enable participants to project and receive positive emotional responses regarding intercultural sensitivity, the most plausible explanation for the difficulty is that the differences of thinking patterns and expression styles among participants may affect their perception on e-mail debate and the improvement of intercultural sensitivity. Thinking patterns refer to forms of reasoning and approaches we use to resolve a problem. Thinking patterns vary from culture to culture and affect the way we communicate with each other. For example, Porter and Samovar (1982) indicated that Western people tend to emphasize logic and rationality by believing that the process of discovering truth follows a logical sequence. In contrast, people in the East believe that the truth will make itself apparent without using any logical consideration or rationale. Kaplan (1966) argued that differences of thinking patterns are reflected in five different language systems: English, Semitic, Oriental, Romance, and Russian. He indicated that the thinking patterns of English speakers are dominantly linear in the language sequence; the Semitic languages (like Arabic) are characterized by a more intuitive and affective reasoning; the Oriental thinking patterns (especially Chinese and Korean) are marked by the writing approach of "indirection"; the Romance languages (e.g., French and Spanish) are the same as English in that they predominantly follow the subject-verb-object order in sentence arrangement, but they allow a greater freedom to digress or to introduce extraneous materials into the conversation; and the structure of Russian language is "made up of a series of presumably parallel constructions and a number of subordinate structures" that are often irrelevant to the central statement (p. 13). The reasoning pattern behind the Russian language is similar to the method of deduction, but it is a time consuming process to reach a result of conversation.

In addition, Ishii (1982) also pointed out that a different reasoning process exists between Westerners and non-Westerners. The former has a linear thinking pattern that leads people to shift from information already stated to information about to be given in order to let the listener understand the speaker, while the latter tend to jump from idea to idea stating the important points without paying attention to the details. The differences in thinking patterns often cause misunderstanding in the process of intercultural communication, especially in the international e-mail exchanges where face-to-face interaction is lacking.

The way we express ourselves is one of the major barriers in intercultural communication (Chen & Starosta, 1998). Expression styles are the spoken performance in the process of expressing ourselves. Our expression styles not only reflect but also embody our beliefs and worldview. Thus, people from different cultures will show different expression styles that can be separated into direct and indirect patterns. Gudykunst and Ting-Toomey (1988) indicated that the direct expression style refers to the use of messages to show one's intentions in interactions and the indirect verbal expression style refers to the use of messages to camouflage and conceal one's true intentions. Hall (1976) pointed out that people in low-context cultures tend to use the direct verbal expression style that emphasizes the situational context, carries important information in explicit verbal messages, values self-expression, verbal fluency, and eloquent speech, and leads people to express their opinions directly and intends to persuade others to accept their viewpoints. In contrast, people in high-context cultures tend to use the indirect verbal expression style that deemphasizes verbal messages, carries important information in contextual cues (e.g., place, time, situation, and relationship), values harmony with the tendency of using ambiguous language and keeping silent in interactions, and leads people to talk around the point and avoid saying "no" directly to others.

The differences of thinking patterns and expression styles between high-context and low-context cultures may also affect the way people communicate via the use of technology. Chung (1992) proposed some potential influences of these differences on e-mail communication between the two groups of people. For example, people from high-context cultures will feel less satisfied with e-mail's function in transmitting relationship building or maintenance information at the initial stage of relationship development but will turn to be more satisfied at the latter stage. Moreover, people from high-context cultures will also feel less satisfied with e-mail's function in expressing feelings toward their partners. Nevertheless, people from low-context cultures will feel more satisfied with e-mail's function in describing details and communicating with logic.

The discussions clearly indicate that culture played an important role in the international e-mail project. The differences of thinking patterns and expression styles dictate the way participants perceive and utilize e-mail communication. Applied to this project, because it was designed as a highly structured debate form for e-mail communication, the format immediately caused orientation problems for some of the participants. "Debate" itself is a product of low-context culture that requires a direct expression of one's argument by using logical reasoning. American, Danish, and German students participating in the project did not show any difficulty in conducting the e-mail debate, while students in France, Hong Kong, and Turkey were confused by the format. The confusion led to two outcomes. First, they resisted or were reluctant to conduct the communication. Second, when they were required to conduct the e-mail debate, they tried to match their American counterparts by abandoning their own expression styles.

To improve the format problem of the global e-mail communication project, a format suitable for both high- and low-context cultures should be designed. We suggest that a regular exchange of information regarding one or several course related topics could be used to replace the debate format. Students in the project should be encouraged to freely share their ideas and opinions from their cultural perspective without being confined by the rigid debate form. Before exchanging information about the specific topic, students should be allowed to use the first two weeks to informally introduce themselves to their counterparts via e-mail. The expressions of feelings and logic can be integrated in this way. In addition, in the process of exchanging information about the specific topic, the instructor can ask students to keep a weekly or bi-weekly journal recording their feelings and opinions toward the project. This process will yield more valid data that are free from the limitation of the format.

As to the improvement of intercultural sensitivity, in addition to the influence of cultural differences, a plausible explanation, as indicated above, is that the one-semester time frame was not long enough. A long-term e-mail interaction may resolve this problem, but it seems not realistic due to the structure of scheduling. Thus, we suggest that for future international e-mail projects it will be more reasonable to assess intercultural awareness rather than intercultural sensitivity, because the cognitive understanding of each other's cultural differences and similarities tend to be more attainable in the limited time span of a project.

According to Triandis (1977), intercultural awareness refers to the understanding of cultural conventions that affect how people think and behave. It requires individuals to understand that, from their own cultural perspective, they are a cultural being and use this understanding as a foundation to further figure out the distinct characteristics of other cultures in order to effectively interpret the behavior of others in intercultural interaction. Because cultures possess different thinking and expression patterns, misunderstanding these differences often causes problems in intercultural communication (Glenn & Glenn, 1981; Oliver, 1962). Thus, to be competent in intercultural interaction, we must first show the ability of intercultural awareness by learning the similarities and differences of each other's culture (Chen & Starosta, 1996). In other words, intercultural awareness is the prerequisite of intercultural sensitivity, and both abilities will in turn lead to intercultural competence (Chen & Starosta, 1997, 1999).

CONCLUSION

The e-mail communication system has become a powerful learning tool for students in the classroom. To apply the e-mail system to a global perspective, a series of assessments and effects deserve further examination. This chapter reports the impact of e-mail exchange on intercultural communication. Problems encountered in this project and suggestions for future experiments in global e-mail exchanges are discussed. Because the trend of global interdependence has created an ever-shifting cultural, economic, ecological, and technological reality that defines the shrinking world of the 21st, we must learn to see through the eyes and minds of people from different cultures, and further develop a global mindset to live meaningfully and productively. In other words, intercultural communication ability has become an indispensable element for reaching the goal of global mindset. To successfully use new technology systems like e-mail for improving students' abilities in intercultural communication will prove to be a promising step toward achieving this goal.

NOTE

This global communication project was supported by a FIPSE grant. Chai Kim was the principal investigator of the grant project on which this chapter is based. Guo-Ming Chen, Linda Shamoon, and Stephen Wood were the co-directors of the project. The original copy of this chapter was previously published in a special issue of *The Edge: The E-Journal of Intercultural Communication* on Intercultural Relations and the Internet (www.hart-li.com/biz/theedge). Reprinted here with permission. Changes have been made for the purpose of this book.

REFERENCES

Adler, N. J. (1997). *International dimensions of organizational behavior*. Cincinnati, OH;: South Western.

Belay, G. (1993). Toward a paradigm shift for intercultural and international communication: New research directions. In S. A. Deetz (Ed*.), Communication Yearbook 16* (pp. 437–457). Newbury Park, CA: Sage.

Chen, G. M. (1993). Intercultural communication education: A classroom case. *Speech Communication Annual, 7,* 33–46.

——. (1994, November). *E-mail debate as a learning tool: A classroom case.* Paper presented at the annual meeting of Speech Communication Association, New Orleans, Louisiana.

Chen, G. M., & Kim, C. (1993, May). *Global classroom: Internationalizing the business curriculum via e-mail.* Paper presented at the Global Classroom Assessment Conference. Braunschweig, Germany.

Chen, G. M., & Starosta, W. J. (1996). Intercultural communication competence: A synthesis. *Communication Yearbook 19* (pp. 353–383). Newsbury Park, CA: Sage.

——. (1997). A review of the concept of intercultural sensitivity. *Human Communication, 1,* 1–16.

——. (1998). *Foundations of intercultural communication.* Boston, MA: Allyn & Bacon.

——. (1999). A review of the concept of intercultural awareness. *Human Communication, 3,* 27-54.

Chen, G. M., & Wood, S. (1994). E-mail debate as a tool of learning. *Speech Communication Teacher, 9,* 15–16.

Chung, J. (1992, November). *Electronic mail usage in low-context and high-context_cultures.* Paper presented at the annual meeting of Speech Communication Association, Chicago, Illinois.

Cohen, M., & Miyake, N. (1986). A worldwide intercultural network: Exploring electronic messaging for instruction. *Instructional Science, 13,* 257–273.

Collier, M. J., & Thomas, M. (1988). Cultural identity: An interpretive perspective. In Y. Y. Kim & W. B. Gudykunst (Eds.), *Theories in intercultural communication* (pp. 99–120). Newbury Park, CA: Sage.

Cooper, M. M., & Selfe, C. L. (1990). Computer conferences and learning: Authority, resistance, and internally persuasive discourse. *College English, 52,* 847–869.

December, J. (1996). Units of analysis for internet communication. *Journal of communication, 46,* 14–38.

Dern, D. P. (1992). Applying the Internet. *Byte, 17,* 111–118.

Garnsey, R., & Garton, A. (1992, September). *Pactok: Asia Pacific electro-media gets earthed.* Paper presented via the Adult Open Learning Information Network Conference, Australia.

Glenn, E. S., & Glenn, C. G. (1981). *Man and mankind: Conflict and communication between cultures.* New Jersey: Norwood.

Gudykunst, W. B., & Ting-Toomey, S. (1988). *Culture and interpersonal communication.* Newbury Park, CA: Sage.

Hall, E. T. (1976). *Beyond Culture.* Garden City, NY: Anchor.

Harasim, L. M. (1993). Global networks: An introduction. In L. M. Harasim (Ed.), *Global networks: Computers and international communication* (pp. 143–151). Cambridge, MA: The MIT Press.

Hawisher, G. E., & Selfe, C. L. (1991). The rhetoric of technology and electronic writing class. *College Composition and Communication, 42*, 55–65.

Ishii, S. (1982). Thought patterns as modes of rhetoric: The United States and Japan. *Communication, 11*.

Kaplan, R. B. (1966). Cultural thought pattern in inter-cultural education. *Language Learning, 16*, 1–20.

Ma, R. (1993, June). *Computer-mediated conversations as a new dimension of intercultural communication between college students in Taiwan and USA*. Paper presented at the International Conference of Chinese Communication Research and Education, Taipei, Taiwan.

——. (1994). Computer-mediated conversations as a new dimension of intercultural communication between East Asian and North American college students. In S. Herring (Ed.), *Computer-mediated communication*. Amsterdam, Netherlands: John Benjamins.

——. (1998, June). *Internet as a global town square.* Paper presented at the 1998 convention of the Chinese Communication Society, Taipei, Taiwan.

Mediamark Research, Inc. (1998). *Cyber States*. [On-line]. Available at HTTP://www.mediamark.com.

Metes, G., Gundry, J., & Bradish, P. (1998). *Agile networking: Competing through the internet and intranets.* Upper Saddle River, NJ: Prentice Hall.

Nua Internet Surveys (1998). *Recent Internet Statistics*. [On line]. Available at HTTP://www.nua.ie.

Oliver, R. T. (1962). *Culture and communication: The problem of penetrating national and cultural boundaries*. Springfield, IL: Thomas.

Porter, R. E., & Samovar, L. (1982). Approaching intercultural communication. In L. A. Samovar & R. E. Porter (Eds.), *Intercultural communication: A reader* (pp. 26–42). Belmont, CA: Wadsworth.

——. (1994). An introduction to intercultural communication. In L. A. Samovar & R. E. Porter (Eds.), *Intercultural communication: A reader* (pp. 4–25). Belmont, CA: Wadsworth.

Shamoon, L. K. (1998). International e-mail debate. In D. Reiss, D. Selfe, & A. Young (Eds.), *Electronic communication across the curriculum* (pp. 151–161). Urbana, IL: National Council of Teachers of English.

Stewart, E. C. (1979). Outline of intercultural communication. In F. L. Casmir (Ed.), *Intercultural and international communication* (pp. 265–344). University Press of America.

Thompson, D. P. (1987). Teaching writing on a local area network. *T.H.E. Journal, 15*, 92–97.

Ting-Toomey, S. (1988). Intercultural conflict style: A face-negotiation theory. In Y. Y. Kim & W. B. Gudykunst (Eds.), *Theories in intercultural communication* (pp. 213–235). Newbury Park, CA: Sage.

——. (1989). Identity and interpersonal bond. In M. K Asante & W. B. Gudykunst (Eds), *Handbook of international and intercultural communication* (pp. 351–173). Newbury Park: Sage.

Triandis, H. C. (1977). Subjective culture and interpersonal relations across cultures. In L. Loeb-Adler (Ed.), Issues in cross-cultural research. *Annals of the New York Academy of Sciences, 285*, 418–434.

CHAPTER NINE
Globalization, Computer-Mediated Interaction, and Symbolic Convergence

Gary W. Larson
Wichita State University

INTRODUCTION

One can hardly address the topic of communication and globalization without critically examining some of the implicit warrants behind this deceptively simple phrase. Communication and globalization imply a sharing of communicative structure across a wide range of cultural, ethnic, and socioeconomic strata as well as across the full geopolitical spectrum. The phrase implies that people may be moving toward a broad base of common understanding—if not sharing—of an homogenized ethos. Globalization, in this regard, is a double-edged sword: on one hand, the concept allows for the possibility of the reduction of communication and communicative process to a "lowest-common-denominator" status in which everyone knows what everyone else is saying because of the lack of depth of the communication; on the other hand, globalization recognizes that more diverse segments of the global community are now able to take part in ongoing discussions about lived realities.

The Internet provides researchers an arena in which to view the enactment of—and resistance to—a global dialogue. While the World Wide Web has become little more than a commercially-colonized *agora* in which the new communicative paradigm is exemplified by IBM's recent U.S. television campaign for doing business on the Web, and Usenet discussion groups often resemble cyber-Levittowns in their cross-channel sameness, there is a facet of Internet communication that retains a dynamic and at times even anarchistic character: the synchronous discussion channels known as Internet Relay Chat. There are other facets of the Internet that also are synchronous, such as ICQ and CUSeeMe (and their respective cousins, America Online's Instant

Messaging and Microsoft's Net Meeting), but Internet Relay Chat—or IRC—still attracts more participants than either. This chapter describes the instrumentality and community of IRC, and demonstrates how a venerable theoretical perspective might be extended for use in studying this medium of social exchange. If the observation of more global communication is accurate, a deeper understanding of just how people make themselves understood as they cross national, cultural and political boundaries is necessary.

One of the most frequently asked questions about studying the net is: In what way does this technology make a difference compared to other mediated venues? In other words, many people seem not to realize the widespread use of the global Internet for the exchange of information and ideas. The Graphic, Visualization and Utilization Center at the Georgia Institute of Technology (GVU) has been conducting on-line surveys of Internet users for a number of years. What their most recent survey (Spring 1998) showed was a continuation of trends they had already reported: the age of users is rising, while the income and educational levels of users is falling. Many more women are joining the Internet as active participants, particularly in the United States. As more people join Internet communities they receive more of their information from this vast resource and, more importantly, more of their socialization.[1] Other reporting services also show Internet usage to be growing at a rapid rate. Nua Internet Surveys (1998) estimates 147 million users of the Internet, with the largest blocks of users coming from North America (87 million), Europe (32.8 million) and Asia (24.3 million). Cyber Stats by Mediamark Research (1998) estimates over 53 million adults in the United States are Internet users, up 23% from a survey done six months earlier. Roper Starch Worldwide, a commerce-oriented reporting firm, notes that 23% of all Americans have accessed the World Wide Web (1998), and that 52% of online users see the computer as a "tool for empowerment."

One will note that "computer-mediated communication" or CMC is not used in this chapter; rather, the narrower implications of "computer-mediated interaction will be addressed." The difference is slight, but significant. CMC implies a reliance on traditional one-way interpretations of mass media. The chosen shift in terminology, ermphasizes the conversational aspects of chat channels. The interaction is what is important; as Chesebro and Bertelsen (1996) point out: "A shift from content orientation to a concern for form or medium—with an emphasis on image, strategy, and patterns of discourse—has been identified as a central feature of the information age" (p. 23).

Studying the "image, strategy and patterns of discourse" on IRC chat channels is important as there are more and more people using these channels of synchronous discourse[2] for a variety of purposes. Some of the increase in use is attributable to software that makes using the channels easier; some new users are logging onto the channels because they are finding something there they can't find anywhere else, such as enacted "safe" spaces in which to talk. No one really knows exactly how many different people have used the channels or are using them every day, since the three major IRC networks only count instances of connection.

For example, when logging onto an IRC network a user registers a nickname[3] with the server but the nicknames of participants can be fluid—changeable at the whim of the participant. It takes only a few keystrokes to change the nickname by which users are known, the only operational caveat is that two participants cannot have the same nickname at the same time. The ease of changing nickname identity is slightly offset by the user's Internet Protocol (IP) address being attached to the name but the IP is not on the screen at all times, a feature of the architecture of IRC that allows relative anonymity (Reid, 1991). Indeed, anonymity continues to be one of the hallmarks of much Internet interaction and is not limited to individual identifiers such as real names, but also includes cultural identifiers such as race, age, gender, class, and size. The anonymous nature of Internet interaction is the springboard from which strategic performances are enacted.

The point is that although the three IRC networks and all their related servers can handle thousands of users at any one time, those users can be changing as fast as names can be typed in. Since servers are located all over the world there is activity on IRC at all hours of the day and night. The numbers may not be as big as the Web or news groups in any one instance, but the potential for cumulative numbers of regular users is quite high. As the 21st century draws closer, many people are changing the way they present themselves to a wider world. Our strategies of performance shift as fast as the media that enable those strategies. So, while it is important to study the instrumentality of the medium, it is also important that we engage the strategic use of the medium by participants. One way to do this is to extend current methods of analysis to this new medium. Symbolic Convergence Theory (SCT) with its attendant method of Fantasy Theme Analysis (FTA) provides a stable methodological base for such an inquiry. The dramatistic metaphor used in FTA is highly developed and lends itself well to an engagement with performative strategies, and to an extension into new modalities of interaction.

The following sections will first describe Internet Relay Chat (IRC) and FTA. Examples of the application of this theory and method are provided to assist in clarifying our understanding of computer-mediated interaction in a contemporary global setting, in particular two forms of discursive manipulation of discourse, referred to as themes of identity and themes of voice, through the strategic use of IRC software. It is suggested that themes of voice and themes of identity are strategic; in other words, they are planned out in advance in general ways by actor/rhetors and are enacted online only by consciously invoking them.

INTERNET RELAY CHAT (IRC)

Australian graduate student Elizabeth Reid wrote a history of the IRC in 1991 and it was made widely available on-line where it spread quickly around the Internet; it is widely quoted by scholars and non-scholars alike. The original IRC program was written in Finland in 1988 and by November of that year had spread across the Internet (Reid, 1991). IRC is a multi-user synchronous communication program that is available in various forms all over the world. In 1990 the Eris Free network (EFnet) was formed with about 95 servers and became the base of the IRC (Mirashi, 1993). IRC continued to grow until the EFnet couldn't handle all the traffic. In 1993, both the DALnet and the Undernet were formed with smaller numbers of servers and users. These three networks—EFnet, DALnet and Undernet—comprise most of the IRC traffic today. There are some other, much smaller, networks that handle some commercial or specialized traffic, but virtually all individual traffic is on one of the three major server networks.

IRC stands apart from the other aspects of the Internet in its ability to allow users to speak in a synchronous fashion (even in groups or channels of a hundred people or more). Coupled with a higher degree of anonymity than is found in other aspects of the Net, IRC provides a platform for users to speak up about issues with more confidence than they might have in a traditional interpersonal setting. Reid (1991) says that users of IRC treat the medium as a frontier world: "a virtual reality of virtual freedom. Participants feel free to act out their fantasies, to challenge social norms" (p. 16). Howard Rheingold (1993), an early pioneer of communication mediated by computers, echoes Reid's characterization, saying that "IRC is what you get when you strip away everything that normally allows people to understand the unspoken shared assumptions that surround and support their communication" (p. 178). Rheingold becomes effusive about the possibilities engendered by the synchronous aspect of IRC because of his early and continuing involvement with an IRC prototype called "the WELL" (Whole Earth Lectronic Link), calling IRC the "great good place" of the Internet (p. 178).

One might think that a conversation is a conversation, no matter where it is held. With face-to-face communication that may be true enough (although I think the exigencies and environments of conversations are germane to their study), but conversations held synchronously over an electronic medium pose new problems for researchers to tackle. An example of a difference between face-to-face interpersonal communication and interpersonal communication on IRC may be found in how conversations flow. In face-to-face exchanges the conversations are initiated, end, are interrupted and are marked by turn-taking. On IRC many of those same features are textually apparent but out of the control of the actors. While participants may control initiation and breaking, the rest become functions of what is referred to as "lag." Lag can be defined as the length of time it takes a message to go from one participant's keyboard to another's computer screen. Along the way that message may be routed through any number of servers as switching software and hardware[4] plan the most efficient route. Because the amount of lag depends on many factors out of the control of the actors (e.g., the number of people on-line at any one time, the amount of data being sent through the packet switchers, the weather in Topeka, and so forth) it is very normal for people in discussions to "walk" over others' lines, speak "out of turn" or even experience long silences of several minutes between responses.

FANTASY THEME ANALYSIS

Since the early 1970s, Symbolic Convergence Theory and Fantasy Theme Analysis have been used, criticized, and reused; withstanding the tests of time and criticism speaks well of this dynamic and multi-faceted methodology. The theory was the product of the so-called "Minneapolis Circle" under the guidance of Ernest Bormann. These students and colleagues tested the theory, made changes to it, extended it into various aspects of communication, and used it in various ways to extend the boundaries of communication research. A few of the more widely know contemporary studies have examined the rhetorical vision of "The Day After" (Foss & Littlejohn, 1986); television coverage of the Iranian hostage release and the Reagan inaugural (Bormann, 1982a); the "Big Book" of Alcoholics Anonymous (Ford, 1989); the use of the theory in corporate strategic planning (Cragan & Shields, 1992); and the rhetoric of abolition (Bormann, 1985, with a response from Smith & Windes in 1993, and a rejoinder from Bormann in 1995).

The on-line rhetoric of IRC presents a dynamic field in which to extend the critical method of FTA. The dramatistic elements of FTA (i.e., actors, scene, script, sanctioning agent, plot) find an easy fit on IRC as participants enact small dramas each time they join a channel. Because of the unique nature of IRC interaction, however, an extension of the method is needed that takes into account strategies of identity and voice. Identity and voice are manipulable elements on IRC, largely due to the anonymity the medium affords users, and to the non-hierarchical structure of the network. "Identity" refers to those strategies used by participants to create recognizable online characters (characters that may change as participants move from channel to channel in one IRC session). "Voice" means the strategic positioning of participants as they use the capabilities of the medium and its software interface. On IRC, each participant is given the agency to create and maintain an identity—or identities—that may or may not have anything to do with their physicality: those features such as gender, race, age or size that may inhibit communication in other venues. The agency granted to IRC participants comes in two very broad forms: the ability to create online personas, and the ability to control the format of their speech to make it more or less noticeable.

In the language of Fantasy Theme Analysis and Symbolic Convergence, the abilities to create and maintain identity and to control speech format may be understood to be "sanctioning agents"—elements that make speech events possible and viable. A sanctioning agent is anything that impacts the flow of speech, making a free flow of thoughts and ideas possible. In this study, the strategic use of an interactive network environment and the performative use of a software interface to the global network are the sanctioning agents of most interest. Sanctioning agents, of course, are only one of the dramatistic elements of symbolic convergence, as envisioned by Ernest Bormann. The other elements of his dramatistic paradigm—actor, plot, script, and scene—are certainly germane to any analysis of rhetorical acts. But the element of particular interest in the analysis of online communicative behavior is the sanctioning agent. There is an analog to theatrical production that became increasingly apparent from watching various IRC groups form, change, move from channel to channel and disband: while the text on screen provides information to other actors within the channels, how each user enacts their persona and how they position their text against each other were the key markers in a shared understanding... a symbolic convergence.

When a theatrical production is "cast," actors are selected to play roles in the production. Actors may be cast because they physically "look the part," because of their cultural or ethnic or background, because of their experience on the stage, or because of their performances in open auditions. The salient point here is that stage actors are cast into roles by people other than themselves. On IRC, actor-rhetors cast *themselves* in various roles, exercising a degree of agency over on-line realities. It is not unusual for one person who has an active on-line life to be known by various nicknames to various groups. Nicknames carry the "force of fantasy" (Bormann, 1982b) by signifying identity within the label by which one chooses to be called. As actors select nicknames for themselves they participate in an elemental act of identity creation: that of picking or constructing a name that will define them individually, and as members of a cultural collective—a rhetorical community. As Oboler (1995) indicates, "ethnic labels, like all names, are by their very nature abstractions of a reality" (p. xv).

Making a choice to be anonymous or invisible to other actors while on-line is also a strategic theme of identity. IRC actor/rhetors may make themselves invisible while on-line by invoking a relatively simple command. On-line invisibility allows actors to enter channels without their nicknames showing up in the on-screen participants list. An actor might cast herself as invisible to gain entrance to a channel forbidden her (e.g., if it were a "males only" channel), but must then stay silent while on the channel. As soon as an utterance is made into the conversation, the invisibility would be a moot point, as the utterance is labeled with the actor's nickname. In practice, though, many actors seem to feel this level of anonymity is self-defeating: in the initial screens from IRC servers once an actor logs onto a network, a count is made of users and how many of them are invisible. This number is generally very low; for example when in one recent session there were more than a thousand clients on that server, but only eighteen were invisible.

People also make strategic decisions about how they will "position" their speech within the discourse; what mode of voice will be used. They may choose to express all emotion with graphic emoticons (the sideways-typed "faces" that express emotion, such as ":-)" for a smile), distancing themselves from actually typing in the words that would express the emotions being felt. They may choose to speak in the third-person or through scripted "popups" ("popups" are prescripted textual routines that may be invoked by a simple mouse-click), thinking these strategies will remove them from the involvement inherent in first-person speech and give them more control.

Moving about the "stage" of IRC channels, positioning speech against that of others, forces participants to make use of the structural elements of IRC in order to make their voices heard. Since so much of the actual content on IRC channels is mundane, this venue is well-suited for the analysis of, as Chesebro and Bertelsen (1996) call it: "technology as text." The authors say that "technology as text shifts attention from the content of communication technology to the structural features of the technology itself" (p. 75), and this type of approach by the analyst emphasizes the instrumental, interpretive, and reflexive functions of a technology. By treating the technology of IRC as a text in this chapter we can, perhaps, get a better understanding of how people are using the global Internet to experiment with representation and agency in their communicative behavior.

METHOD

The data used in this chapter is not representative of all conversations on IRC. Ethnographic in method, this study sought to describe and analyze communicative behavior on racialized IRC channels. To that end, connections were made to both African-American channels and to white supremacist channels on IRC during varying parts of the day and night, specifically targeting evenings and weekends, Central Standard Time, to try and get the widest possible discussion from participants in the United States. While on the channels the author was a passive observer; not engaging in the on-line discussions of race at all.

The channels monitored were selected from their titles and from their topic lines (e.g. "#blackpower: discussions of interest for African-Americans"). There may have been other channels devoted to racialized discourse but which were listed with non-racial titles or topics. The white supremacist were found in channels the same way: through their titles and their topic lines.

The channels were monitored for at least an hour at a time, using the traffic in the channel as an indicator. If traffic was heavy—in other words, if many people came and went or there was a great deal of "small talk"—monitoring continued for several hours. The author logged-on to the channels when there were scheduled discussions, but at other times as well, amassing over 250 hours of time on the various channels. A single nickname was used in the African-American channels, and a different nickname in the other channels; thus being allowed some freedom to "lurk"[5] without the suspicion it may have aroused had the same nick been used throughout the study.

A great deal of conversational data were collected during the four-month period of this study, and two actions seem to exemplify the strategic, performative nature of on-line communication within IRC channels: "blowing the cover" of someone while on-line and "kicking" someone from a channel. The following data and discussion utilize excerpts of data collected for a larger project; no attempt has been made to sanitize the language in the following excerpts. The language, at times, is quite raw, xenophobic, and racist.

THEMES OF STRATEGIC IDENTITY

When a person walks into a party and wants to join in a conversation, a number of cues help her assess the situation and to achieve a "fit" with what is going on, i.e., being able to see who people are talking to, sampling bits of simultaneous conversation, judging the depth of conversation by the nonverbal cues being offered and so forth. Much of this same type of analysis is available to on-line actor/rhetors on IRC, but is not apparent in a single glance—it must be sought. For instance, it is possible to see in what other groups another person on the channel is speaking[6] by invoking a software command called "whois." By invoking this command, an actor may find out in what other channels another actor is speaking, the e-mail name (but not address) of that actor, and the name that actor has entered as a "true" name. In general, the result of a "whois" command will look something like this:

Gary43 is lars0300@160.94.62.187 * Gary W. Larson *
Signed on to St-Louis.Dal.Net * In channels #BlackEssence,
#BlackPower, #40sAndChatting.

This tells anyone who invokes the command that Gary W. Larson (the author) has an e-mail name of lars0300, an Internet Protocol address (which is not the same as e-mail) of 160.94.62.187, and that he is logged onto three channels at the time of the "whois" command: BlackEssence, BlackPower, and 40sAndChatting (the "#" sign signifies a channel name on IRC).

Invoking the "whois" command becomes strategic when actor/rhetors use it either to cloak themselves in a deeper level of anonymity, or to "out" another actor who is not observing the etiquette of the channel. The latter is more interesting, as this exchange illustrates:

10 BigTom: any supremacist is pretty much a loser... can't we all just make love and get a long ... hehehehe :)
13 *** Nazi has joined #black-3
14 Bob: Nazi :TONY
15 Bob: Nazi :#black-3 #skinhead #sweden #sverige

16 Bob: LOL
17 Nazi: you're nickt
18 *** Nazi has left #black-3

In this exchange, two males are talking about white supremacists taking over other channels. BigTom has just made a statement about that topic when the channel is joined by a user who calls himself Nazi. Bob does a "whois" command and publishes the results in the open channel, effectively outing Nazi as a man named Tony who is also speaking in three other channels, at least two of which, "skinhead" and "sverige" are devoted to racist speech. Apparently flustered by his blown cover, Nazi chooses to leave the channel, a reaction that was apparently intended by Bob all along.

Bob has made strategic use of an instrumental feature of IRC: the ability to find something out about other actors by invoking software commands. This feature is enabling as it protects users from being blindly harassed, but of course the opposite implication is just as true—that this feature of IRC can be seen as a constraint, making public information an actor would not want to be made public. Consider the following exchange from later in the same conversation:

23 Bob: FatMan: he doesn't understand English. Only hickbonics.
27 Nicky: I understand English perfectly well, thank you.
28 Bob: Nicky: not you!!! Nazi! he is in some nazi channel and does't talk on here
29 BigTom: Nazi:#black-3 @#!!!!!!!!!!!niggers @#sheep on the run #dicklessness #assmunchers Anon

In this exchange, BigTom has taken the strategy Bob used earlier and has chained it out into another use: that of ridiculing someone. BigTom has now cast Nazi into several channels (line 29), none of which actually existed except the first one ("black-3") which was this channel. The strategy of ridiculing of someone in an open channel goes back to the formative days of Usenet newsgroups when participants would engage in "flame wars": heated arguments between participants that often devolved into creative name-calling. It is still quite common to see this same sort of exchange on IRC; rejoinders which call into question the other actor's sexuality, as shown above. "Sheep on the run," "dicklessness" and "assmunchers Anon" all, in effect, cast Nazi in pejorative sexual terms.

Another manipulation of identity quite common on IRC is to re-cast oneself in the middle of a conversation. Often this is a reaction to what is being said in the channel, as illustrated by the following exchange from a different conversation:

20 Kitty be careful flesh

22 Kitty im watching you
23 Fleshed theres always a sour apple in tha bunch
24 Fleshed General will soon own this chat
25 * Kitty looks at flesh and feels a change comming on
26 General flesh what are you talkin about????
27 *** Kitty is now known as Panther
28 Panther GRRRRRRRRRRRRLLL!!!!!!!!!!!!!!!!!!!!!!!

In this exchange, Kitty has become tired of the statements being made by Fleshed and wants to let him know that she, as a channel operator, was not going to let it continue much longer. To give Fleshed warning she invokes a simple command to change her nickname from Kitty to Panther and recasts her persona from docile pet to aggressive killer. This action—recasting oneself in a new persona—also speaks to issues of power and dominance.

Another example of recasting one's persona, although far more uncomfortable to read, is found in an exchange from a Ku Klux Klan channel; several people are chaining out the traditional fantasies of the Klan about African-Americans, while a number of other people are on-channel but silent. (This is known as "lurking" and is not appreciated by many channel operators.) After witnessing the conversation for quite a while, one actor finally breaks his silence:

44 Tude: Damn and I am Black too...Means you probably don't like me???
48 Hawg: Come to one of our rallies, Tude you'll turn white with fear, then black again when we burn your nigger ass!
49 007: hey Tude, can't we all get along?
50 Tude: I love you man!!!
51 *** Tude is now known as Black
52 Black: I'm proud to be black!!!
54 * Warrior Takes out well used nigger whip and beats the shit out of Black "Call it Roots 2"
55 *** Black was kicked by Bot (•[X]• BuckWheat : We must secure the future for our race and our white Children)
56 Reverend: Black is normally in #gayblackmen room.
57 Black: Grow up little men!!!
58 Black: Boys get some new lines...LOSERS HE HE

In this exchange, the actor named Tude had been lurking until he decided that he had to say something. Knowing the type of reaction his pronouncement of black identity would bring, Tude recast himself as Black, a proud African-American. The reaction was predictable: Black was kicked off-channel by Bot[7] after being symbolically whipped by Warrior. As he leaves, Reverend remarks that Black is "normally" in a channel devoted to homosexuals. (It is interesting to note here that no one else on the channel questions how Reverend would know Black is normally on a homosexual channel.) Because he was only kicked and not banned from the channel, Black makes a reappearance to tell the other actors on the channel to grow up and get some new lines. An actor who is kicked from a channel may re-enter that channel immediately. An actor who is banned from a channel, however, may not re-enter that channel unless the channel operator (or "ops") allows it—more on this in the next section, but briefly, any IRC channel may have several ops; that status is granted to the person who starts the channel, and may be passed along to others at that person's pleasure.

Interesting to me in this exchange as well as some others from white supremacist channels is the selection of nicknames, the reliance on the rhetorical vision of homosexuality as aberrant, and the predominant use of "popups" and "bot" responses; the latter being themes of voice.

THEMES OF STRATEGIC VOICE

Positioning one's speech against the speech of all others in a channel makes use of both the constraining as well as enabling instrumental functions of IRC. Strategic methods for finding one's voice can happen in a number of different ways, some of which are dependent upon the version of software an actor is using to gain access to IRC channels. For example, the version of software used in this study was updated about halfway through the period of the study. One change in the software was the inclusion of provisions for changing the color of text on the screens of people who are on-channel.

The major themes of voice illustrated in this section will be those of third-person construction, popups, and bots. Third-person construction is a common way to set speech apart on even the oldest versions of IRC software. By typing "/me..." an actor tells the IRC server to start the next line with the actor's nickname and fill in whatever follows, effectively creating a third-person construction. Popups are scripted utterances and responses that actors may prepare in advance of an online session and save to their own computers. When the IRC software is opened the popup files are enabled, giving the actor access to whatever scripts have been saved at a simple click of a mouse button. Finally, bots are small software subroutine programs that are also saved to an actor's own files and are again enabled when IRC software is started. Bots may be installed in channels and will remain active as long as the person who installs them is online. Briefly, bots usually perform some function on the channel, anything from automatically greeting newcomers to serving cyber-drinks from a cyber-bar!

Third-Person Construction

Third-person construction of online text usually takes the form of some sort of declaration: either describing what the actor is "doing" within the constructed on-line setting, making a statement about the intentions or feelings of the actor, warning another actor of impending action, or self-describing either on-line or off-line characteristics. In the following lines taken from a number of conversations, all these uses may be seen:

* Sunny says hello to all and blows a kiss to all the men.
* Dawg thinks the girl in the movie wuz fine...for an older woman.
* Dethshd is gonna find a channel with a little less hostility in the air.
* Kitty looks at flesh and feels a change comming on.
* Simon takes out the BULLWHIP and tells Flaming "I'm gonna teach you some manners, boy!"
* Executioner is also 100% German, the perfect race, or "Master" race as I call it.

In each case, the declarative line, when typed and entered, appeared on the screens of all other on-line actors without initial bracketing of a nickname and in purple rather than black.

Third-person declaratives are important strategic themes; there must be a cognitive turn from the actor/rhetor to construct the declaration to begin with. We normally don't think of ourselves in the third person and so expressing something about ourselves in third-person requires us to step outside ourselves and to view ourselves as part of a larger narrative, a story. The distancing of self from personalized text allows actor/rhetors to say things they wouldn't normally say in their lived realities, to be bold and courageous on-line when they might be shy and fearful in their lived realities. Third-person constructions are often scripted well in advance of any conversation, seemingly in anticipation of topics. This forethought represents a highly strategic use of the instrumentality of IRC as it implies the actor/rhetor either knows or can guess what topics will be discussed on the channel, or that the actor/rhetor will turn the conversation to a topic for which scripted popups have been written.

Popups

By far the greatest use of popups I saw while monitoring the IRC channels was the use of lengthy, pre-scripted utterances. If the popup invoked does not begin with a command, such as "/me ..." it appears as if the actor has typed it in on the spur of the moment in typical IRC-discursive fashion. What identifies the utterance as a popup in this situation is its length, complexity, and whether or not it is repeated *verbatim*, either by the same actor or another. While the channels populated largely by African-Americans showed some use of popup scripts in this fashion, the channels antithetical to Black channels showed major use of scripted speech popups.

A good example of this is found in the following lengthy fragment; all four pages of the obvious popup scripts are not listed as a couple examples will more than suffice:

1	Reverend	The UnReverend Jesse Jackson quoted from the Dec 17, 1993 edition of the Wall Street Journal: Jesse Jackson said the other day that "there is nothing more painful to me . . . than to walk down the street and hear footsteps and start thinking about robbery—then look around and see somebody white and feel relieved."
2	Language	For 1994, for every 100,000 white male adults, 860 were in jail or prison (white females, 60). For every 100,000 black males, 6,753 were in jail or prison (black females 435) [1995 Sourcebook, table 6.12]. The black male rate is 7.85 times as great as for white males (black female rate is 7.25 times the white female rate).
9	Hunter	For 1993, of all families on AFDC (Aid to Families with Dependent Children), 37.2% were black (but only 12.4% of the population). Whites (most of the population) were only 38.9% of AFDC families. One in 3 black adults on AFDC will still be there in 10 years (versus 1 in 5 for whites). In 1992, 35.1% of all black children were on AFDC. [Source: *NY Times*, 23 March 1995]

This particular fragment of text reflects a strategic use of scripted popups. The fragment was taken from a predominantly black channel that was in the process of being taken over by white supremacists at the time I logged on.

Reverend, Language, and Hunter were the three main instigators of the take-over and had scripted lengthy popups with all kinds of cited "evidence" for their arguments, most of it taken out of any sort of context or from controversial works, such as Herrnstein's *The Bell Curve* (1994). The text in this instance is less important than the performance of the text. Predominantly vacuous, the sheer volume of the text and the speed with which it was generated indicate pre-scripting. The text from the take-over, which constitutes about four pages of conversation, was generated in less than an hour of on-line time. By comparison, the last page of a conversation from a channel to which actors escaped during the take-over, took almost as long to generate, yet contains a fraction of the amount of text.

Bots

Of all the conversations monitored, the following is one of the most difficult to read as it is one of the most hateful of all the discussions encountered. The highly scripted popups of the previous fragment attempted to use the device of a preponderance of evidence, even though the validity of the evidence was suspect. The following fragment, on the other hand, represents a type of racism that doesn't hide behind manipulated statistics or quasi-rational appeals: it simply is fear of the "other":

1 *** Bot sets mode: +o Reverend
5 Bot: nod Im not Racist I like niggers I think everyone should own a few
7 baddog: Help save humanity, lynch a nigger today!
8 Bot: Nigger Nigger Kill the Nigger
9 evilkid: hey any seriouse white people that want to end the rise of all niggers in this country
10 Bot: Nigger Nigger Kill the Nigger
—
29 wimps: there just people to and i am not black smart guy
30 Bot: Black ohh you mean worthless nigger
34 wimps: well if you had a dick you wouldn't know what to do with it
35 Bot: wimps you will hang!!!
38 *** wimps was kicked by Bot What does not kill me will 0nly make me strounger)

These two exchanges from one fragment illustrate some of the uses to which bots are put in on-line settings. In the first exchange, Bot automatically gives power to Reverend in the form of channel operator status on line 1. Bot also responds to a "trigger" word for which an automated textual response has been programmed. In this case, the word "racist" in line 4 brings an automated response in line 5, just as the word "nigger" in line 7 brings the automated response of line 8. In the following two exchanges, Bot responds to the trigger word "black" and in each case the result is the automatic expulsion of an actor/rhetor from the channel (line 38). These are very typical of the uses for on-line bots, but like pre-scripted popups, a far greater use of them was found in white supremacist channels than in African-American channels.

DISCUSSION

The performative themes of identity and voice described in this chapter illustrate what Christopher Werry (1996) calls "prosodic" and "orthographic" strategies. Prosody, according to Webster (Morehead, 1981), is the "science of metrical versification" (p. 424). In other words, prosody involves the colorful use of language in conversation. In IRC chat, prosodic strategies aid actors in creating and maintaining identities while in a channel. When an actor casts herself in a named role by assuming a nickname and by building a persona, it is strategic in the initial acts of planning a nickname to decide what level of anonymity to employ while on-line. The inventive use of a "whois" command to reveal truthful or deceptive information about another actor can be seen as self-defense: if someone is disturbing your on-line community, revealing the "truth" about that actor will bring to a head or even stop the disturbance. Recasting oneself in a different role while in a conversation can also be seen as prosodic strategy. Recasting is almost always a reaction to a prior event in the channel; the subsequent manipulation of identity both confirms the salience of the initiating act and provides a response to it. Identity-manipulation and self-defense may be seen as two prosodic themes that have developed wide use on IRC; they certainly are manifest in the channels monitored in this study.

Orthography, again according to Webster (Morehead, 1981), pertains to correct spelling of words. On IRC, orthographic strategies may be thought of as using the enabling features of the software to format text. In other words, what have been referred to as themes of voice are primarily orthographic in nature; using popups, bots, and colored text help position an actor's voice among the babble of speech on any IRC channel. Third-person construction seems to indicate a desire for distance, either from other actors or even from oneself. By removing oneself to the position of a distanced third person, an actor assumes a quasi-neutral stance which allows her to comment on the action in the channel as if she were an objective observer.

A discursive goal for the use of popups might be seen as a desire for efficiency while on-channel, with the IRC software making it possible to pre-script monologues on a variety of topics and to invoke them with a click of a mouse button. While this is an efficient way to generate great quantities of speech, it does not necessarily have the same impact on quality; witness the extensive use of popups on the white supremacist channels with large quantities of vacuous and often misspelled content. Quantity, in this instance, was far more important to the originating actors than quality; it was the sheer quantity of speech that allowed the actors to force others off the channel, effectively taking over the channel by the use of voice.

The linguistic concepts of prosody and orthography, when extended to conversation analysis on IRC, reveal distinctly pragmatic avenues of analysis for the performed nature of identity and agency of online actors. Combining the concepts with the Fantasy Theme element of sanctioning agent gives the analyst a way of extending existing theory into a new modality of communicative behavior. The examples used in this chapter clearly show a pattern of strategic manipulation of both identity and voice through the manipulation of the network environment and of the enabling software. The ease with which prosodic and orthographic elements can be manipulated demonstrates how the interactive environment of the network and the performative environment of the software are sanctioning agents for computer mediated interaction. In the case of the larger study from which this chapter was drawn, this methodology allowed for some inference about polarized online communities; about how online communities often reclaim the symbolic territories that allow colonialized lived realities—in the case of the larger study, the lived reality of a racialized and often racist physical world.

While it is dangerous to generalize from an ethnographic account of activities, the instrumentality of IRC demonstrably enables participants to dynamically manipulate their perceived identities to a degree not possible in other mediated contexts like television, newspaper or the World Wide Web. It allows people to experiment with new personae, to see how a person with a different cultural, racial, or ethnic background would perceive or be perceived in the same situations.

Bormann, Cragan, and Shields (1996) demonstrated that consciousness-creating communication generates new symbolic ground for a community according to the principles of novelty, explanatory power, and imitation. The principle of novelty asserts that rhetoricians must be innovative to keep rhetorical visions from lagging behind contemporary conditions. The principle of explanatory power says that people are likely to share fantasies that account for or make sense of a confusing reality; and the principle of imitation tells us that when people are bored or confused they often share fantasies that give old dramas a new face. Rhetorical communities on IRC may be seen demonstrating this phase in the life-cycle of rhetorical visions: a community consciousness is created by taking the history of the rhetorical community and re-framing it for a contemporary modality of communication. The prosodic and orthographic themes of identity and voice are enacted by participants in these rhetorical communities to perhaps give participants a sense of power and control that may have been lacking in their physical worlds. In this sense, the performative tropes outlined in this chapter liberate actors from their physical worlds, allowing them at least a sense of agency from which to construct their own identities and communities.

CONCLUSION

We return, then, to the issue of globalization. With an estimated U.S. population on the global Internet at approximately 60% (Nua Internet Surveys, 1998), it is small wonder that English has become the *de facto* linguistic standard for communicating in cyberspace. Questions must be asked, however, about the effects of this colonization of virtual space. For example, are there pockets of "resistance" in which no English is allowed to be spoken? The answer is, of course, "yes." On IRC there are many channels in which statements and questions in English are either ignored or are cause for the participant to be ejected from the channel. In these channels, however, a quiet observer may witness the same sort of communicative behavior described in this chapter. The use of third-person construction, popups, bots, name/identity changes—in other words, the same sort of prosodic and orthographic strategies used in English-speaking channels—are prevalent in many disparate channels. One of the tenets of Symbolic Convergence Theory is that fantasy themes "chain out" into rhetorical communities and are used in circumstances other than the original conversation. The widespread use of familiar prosodic and orthographic themes demonstrates the validity of using this method of analysis, centering on strategic and performative themes as sanctioning agents. The dynamic negotiations of identity and agency seem not to be confined to any single demographic stratum or subset of strata; observation, description, and analysis of those ongoing negotiations will grant us important clues about trans-national communication in the future.

Another question that needs to be addressed is global access. As the Nua survey (1998) observes, 98% of all traffic on the global Internet is from either North America, Europe, or Asia. The wide disparity of access between industrial and developing regions of the world threatens to create even wider chasms between information "haves" and "have nots." While questions of technological infrastructure do not fall within the bounds of this chapter nor of the larger study from which it was taken, it would be remiss not to raise the issue. Indeed, even some past celebrants of the Internet have recently been raising questions about the need for this connectedness (Clifford Stoll, 1996). In any case, universal access to global lines of communication is a controversial topic that needs far more study in any discussion of globalization and communication.

NOTES

1. There is no shortage of reports of Internet usage. M.I.T. (Massachusetts Institute of Technology, 1998) mapped early Internet use from the National Science Foundation (who stopped running the Internet backbone in April 1995). Those numbers showed the largest use of the Net came from World Wide Webs users (32.9%), followed by FTP (file transfers, at 24.2%), and distantly followed by Usenet, E-mail, Gopher, and IRC.

2. According to a study done by Mika Nystrom at M.I.T. (Nystrom, 1993, p. 3) in 1993 of users on the oldest of the IRC server networks, EFnet, usage increased from an average of 606 users at any one time in March 1992, to 2,316 in April 1993. According to the study the growth rate is higher than a linear curve, leading Nystrom to call it an "explosive increase."

3. Nicknames are part of the reason so much discussion happens on IRC. A user may certainly use her or his actual name but the great majority of users select nicknames they feel describe some aspect of themselves which may not be visually apparent in a text-only medium of conversation.

4. This is referred to as "packet switching." A packet is a small amount of information encoded with a digital header and address that tells the system where it is supposed to go, how many packets are in the total message and which packet this is of the total. In this way, messages may be broken into small digital "chunks," routed along different routes, and still arrive and be assembled in the proper order in a matter of seconds. It is the systems theory concept of "equifinality" made manifest.

5. "Lurking" is the practice of logging onto channels and staying without saying anything. In some channels this is considered polite behavior, while in others—such as the white supremacist channels—

it is considered suspicious behavior and is grounds for being kicked off the channel. If a person who has been kicked comes back without an explanation that person may be permanently banned from the channel.

6. It is possible on IRC to speak in many groups at once. A user may have open a number of windows on their screen, each of which contains the text/conversation of a different channel, and may be contributing to all of those channels.

7. A "bot" is a generic term for a small software sub-routine that may be installed into a channel. Bots may serve many purposes, from automated greetings when a person joins a channel, to serving symbolic drinks when an actor calls for one, to kicking people off channel either when called to do so by someone with operator status or even automatically when an actor says something that triggers the kick. Bots of the most destructive sort, called "floodbots," read the nicknames of people on-channel and flood their IP address with text to the point that those people are disconnected from the Internet. That type of bot is ostensibly banned from most IRC servers, but in reality still shows up from time to time.

REFERENCES

Bormann, E.G. (1982a). A fantasy theme analysis of the television coverage of the hostage release and the Reagan inaugural. *Quarterly Journal of Speech, 68*, 13–45.

——. (1982b). Fantasy and rhetorical vision: Ten years later. *Quarterly Journal of Speech, 68*, 288–305.

——. (1985). Symbolic convergence theory: A communication formulation. *Journal of Communication, 35*, 128–38.

——. (1995). Some random thoughts on the unity or diversity of the rhetoric of abolition. *Southern Communication Journal, 60*, 266–74.

Bormann, E.G., Cragan, J.F. & Shields, D.C. (1996). An expansion of the rhetorical vision component of the symbolic convergence theory: The cold war paradigm case. *Communication Monographs, 63*, 1–28.

Chesebro, J.W. & Bertelsen, D.A. (1996). *Analyzing media: Communication technologies as symbolic and cognitive systems*. New York: Guilford Press.

Cragan, J.F. & Shields, D.C. (1992). The use of symbolic convergence theory in corporate strategic planning: A case study. *Journal of Applied Communication Research, 20*, 199–218.

Ford, L.A. (1989). Fetching good out of evil in AA: A Bormannean fantasy theme analysis of the Big Book of Alcoholics Anonymous. *Communication Quarterly, 37*, 1–15.

Foss, K.A. & Littlejohn, S.W. (1986). The day after: Rhetorical vision in an ironic frame. *Critical Studies in Mass Communication, 3*, 317–36.

Herrnstein, R.J. (1994). *The bell curve*. New York: Free Press.

Massachusetts Institute of Technology. (1998). *National Science Foundation backbone Internet statistics.* [On-line]. Available at HTTP://www.mit.edu.

Mediamark Research, Inc. (1998). *Cyber Stats*. [On-line]. Available at HTTP://www.mediamark.com.

Mirashi, M. (1993). *The history of the undernet*. [On-line]. Available at HTTP://www2.undernet.org:8080/~cs93jtl/unet_history.txt.

Morehead, A. (Ed.) (1981). *The New American Webster Handy College Dictionary*. New York, NY: Signet.

Nua Internet Surveys. (1998). *Recent Internet statistics.* [On-line]. Available at HTTP://www.nua.ie.

Nystrom, M. (1993). *Recent growth in number of IRC clients and projected future growth.* [On-line]. Available at HTTP:// gopher. it.lut.fi:70/0/net/irc/.

Oboler. S. (1995). *Ethic labels, Latino lives.* Minneapolis, MN: University of Minnesota Press.

Reid, E. (1991). *Electropolis: Communication and community on IRC.* Unpublished honors thesis, University of Melbourne. [On-line]. Available at HTTP://www2.undernet.org:8080/~cs93jtl/Electropolis.

Rheingold, H. (1993). *The virtual community: Homesteading on the electronic frontier*. Reading, MA: Addison-Wesley.

Roper Starch Worldwide. (1998). 1998 *Demographic Profile*. [On-line]. Available at HTTP://www.roper.com.

Smith, R.R. & Windes, R.R. (1993). Symbolic convergence and abolitionism: A terministic reinterpretation. *Southern Communication Journal, 59*, 45–59.

Stoll, C. (1996). *Silicon snake oil: Second thoughts on the information highway.* New York: Anchor.

Werry, C. C. (1996). Linguistic and interactional features of internet relay chat. In S.C. Herring (Ed.), *Computer-mediated communication: Linguistic, social and cross-cultural perspectives* (pp. 47–63). Amsterdam: John Benjamins Publishing Company.

CHAPTER TEN
Ethical Issues in Global Communication

Robert Shuter
Marquette University

INTRODUCTIOIN

The technological revolution has dramatically increased communication across the globe. Messages are being transmitted continually across electronic media to corners of the globe that were considered remote just ten years ago. In addition to telecommunication linkages enhanced by satellite transmissions, the Internet has become a fixture in the world's technological arsenal. With the explosion of electronic messages worldwide, ethical issues have surfaced concerning the use of global communication technologies. Although each communication technology may have a particular set of ethical concerns, this chapter addresses generic ethical issues that affect all communication technologies from telecommunication to the Internet.

As communication increases globally, it is clear that some of the world's citizens have more access to information and new technology than do others. In truth, information elitism is becoming increasingly apparent. New communication technologies diffuse quickly to North America and Europe while Latin America, Africa, and much of Asia receive technology more slowly. Consider telephones, for example. Boafo (1991) notes that black Africa possesses two telephones per 1,000 people while Tokyo has more telephones than all of sub-Saharan Africa. Brazil, Chile, and Mexico have approximately six and India and Indonesia .6 telephones per 100 people while the U.S., United Kingdom, and Germany have over 44 lines per 100 citizens. This huge disparity impacts on all aspects of communication life including phone access, availability of communication networks like the Internet, and connecting in general to the information superhighway.

Satellite communication offers another example of the disparity between technological haves and have-nots. First, Western powers have clearly dominated satellite communication as evidenced by 72% of satellites in 1989 serving the industrialized countries and only 14 satellites available for the rest of the world (Hudson, 1990). Hudson (1990) concludes that satellites clearly benefited middle and upper income communities because they had the resources to lease transponders from Intelstat and, hence, participate in the satellite revolution.

Some have argued that "information domination" has actually increased with the deregulation and privatization of communication industries (Sussman & Lent, 1991). That is, global corporate entities have exploited the hue and cry for deregulation by taking over local communication industries in countries across the world. Sussman and Lent (1991) identify transnational mergers by selected corporations that produced further Western corporate domination of communication industries. These corporate communication giants include Alcatel (France), AT&T (U.S.), GTE (U.S.), Siemens (Germany), and NTT (Japan). Sussman and Lent's (1991) list predates the meteoric rise of Microsoft that has so dominated the global software industry that the U.S. government recently sued Microsoft for monopolistic practices.

More recent data on western corporate control of global communication show Reuters Television (U.K.) and CNN (U.S.A.) leading all other international news agencies; Bertelsam (German) and Walt Disney (U.S.A.) as the largest entertainment media corporations; and WPT (U.K.) and Saatchi and Saatchi (U.K.) first among advertising agencies (Lent, 1991). Japan is the only nonwestern country that has sizable technology companies with Toshiba second only to IBM in computer manufacturing. European and U.S. domination of international communication shifts control of technology, media, and their messages from individual countries to transnational corporate entities which determine access to and content of information. In support of this arrangement, it is argued that only by deregulating government-owned communication industries can market entities like transnational corporations prosper and provide high quality communication services to people worldwide. The ethical tradeoff is clear: relinquishing local autonomy for potentially improved communication services.

INFORMATION IMPERIALISM OR CULTURAL PLURALISM?

With the control of information production and transmission by European and U.S. corporations, cultures worldwide are exposed to communication over the Internet and airways that may not be compatible with local customs, politics, or ethical standards. Consider Brazil, for example. Oliveira (1991) argues that advertising agencies flood Brazilian markets with commercial messages that promote consumption of products produced in North America. Interestingly, many of these commercials focus on consumer items that only a small percentage of Brazilians can afford given the huge disparities in wealth in Brazil. These commercial messages influence cultural tastes and buying patterns and create dependency on North American products (Oliveira, 1991). In 1990, the Brazilian Minister of Culture railed about dependency on North American culture when he said: "The phenomenon of cultural domination has reached unprecedented proportions. We have now entire societies that are basically consumers of culture, and others that are producers of culture" (Oliveira, 1991). Nowhere is this more evident than in television programming in Brazil where over 3/4 of all shows are imported from the U.S. This programming bombards Brazilian viewers with U.S. lifestyles and cultural values that some fear may have an eroding effect on Brazilian culture.

For many in the information business, the charge of cultural imperialism is perceived to be motivated by the desire to restrict people to a steady diet of certain types of information that normally serve the interests of a ruling elite or state controlled ideology. Ethically, people need information choices, so the cultural pluralists argue, and this can only be achieved by diversifying a country's information access and reception. A free and unfettered information superhighway is a liberating force, according to free market advocates, because it exposes the world's citizenry to a plethora of ideas, tastes, traditions, and morals. The tension between information imperialism and cultural pluralism is best exemplified in China.

The information revolution is quickly taking hold in China. Telecommunication has mushroomed with China installing more than 73 million phones since 1973, which is more phone installations than the entire developing world combined (Mueller & Tan, 1997). China has more than 28 million pagers, second only to the United States. And though cell phones are just beginning to emerge in China; it is one of the top five markets in cell phones with over 4.7 million subscribers (Mueller & Tan, 1997). China is knee-deep in new communication technology, and this poses new opportunities and threats to the Chinese government.

The Internet may be the most provocative and threatening new communication technology. While most Chinese are not computer literate, the volume of personal computers in China, which was approximately 3 million in 1996, is growing by 40% a year. Computer literacy is generally reserved for the most educated Chinese who are often business leaders, professors, and graduate students at the leading universities. These information elites generally have more contact with international sources and, hence, rely heavily on the Internet. For the Chinese government, the Internet is a source of potentially threatening ideas which can not be allowed to grow without restrictions (Mueller & Tan, 1997).

The official governmental position is that the Internet, like other international media, poses risks to the economic, cultural, and political well being of the country (Tefft, 1995). To control the Internet, China has done several things including (1) maintaining a government monopoly (MPT) on international telecommunication connections, (2) requiring all international data to go through MPT, and (3) registering all end users. These measures place end users and information sources under the control of the government, which seriously curtails the freedom of the Internet (Mueller & Tan, 1997).

China also protects its communication market by not allowing direct foreign investment in telecommunication. By controlling telecommunication, China maintains a formidable surveillance apparatus. Also, centralization allows China to monopolize the lucrative and growing telecommunication industry. Given the sophistication, size, and resources of foreign corporations, China would not be able to compete successfully with them.

China's policy to control electronic communication is clearly at odds with the U.S. notion of a free and unfettered media. Like China, many countries with a developing communication infrastructure restrict access to their media markets and control media content. At the heart of this debate are conflicting ethical assumptions about freedom, self-expression, and responsibility. The next section of this chapter explores conflicting ethical assumptions that influence western and eastern communication markets.

ETHICAL FOUNDATIONS OF WESTERN COMMUNICATION ETHICS

Communication ethics grounded in a western model presumes that information, regardless of its controversy, should be available to listeners and media/new communication technologies should be able to pursue truth without significant control by external forces. The first amendment, from a U.S. perspective, provides a constitutional guarantee for a marketplace of ideas which also protects electronic media from undue government control. This approach to communication ethics flows from a western perspective that human beings should choose their course of action and can be judged right or wrong on the basis of those choices.

Wargo (1990) indicates that the ascendance of the individual in western society springs from a Judeo Christian religious perspective that bifurcates the relationship between God, people, and nature. An implicit hierarchy is apparent in the Judeo Christian view that places God at the apex followed by human beings and lastly nature. It is the human soul—the spiritual side of being—that gives rise to ethical judgments.

Inherent in the Judeo Christian view of ethics, according to Wargo (1990), is free will—the presumption that an intellect, which only human beings possess, provides individuals with the ability to make choices. To sin, then, is to choose a course of action incompatible with divine law. The Judeo Christian emphasis on free will and choice are embedded in western communication ethics.

Nilsen (1966) writes that for communication to be ethical, choice must be based on the best information available. Access to information is key to developing ethical messages because it provides listeners with sufficient facts to make informed choices. It is not surprising that countries grounded in a Judeo Christian perspective often have developed codes of conduct for media professionals to protect information access and journalistic freedom (Cooper, Christians, Plude, & White, 1989). Also, the emphasis on reason and logic—so central to a Judeo Christian perspective—is also reflected in communication policies and practices in the U.S., which establishes reason and logic as touchstones for determining the ethical nature of messages.

For Western communication theorists, reason and logic are critical dimensions of ethical messages. Aristotelian philosophy emphasizes reason and logic, and reduces the value of emotional appeal (Shuter, in press-b). Since reason is central to free will, ethical discourse emphasizes logical arguments so that listeners can make reasoned choices when exposed to a message.

Finally, communication ethics in the U.S. and the West requires that all listeners have equal access to information, regardless of race, social class, religion, or national origin (Shuter, in press-a). Broadcast entities that discriminate on the basis of race, sex, or national origin can lose their media licenses and, hence, are required by law to make information as universally available as possible. Messages that demean or injure a person's race, ethnicity, gender, religion, sexual orientation, or national origin are also considered unethical.

The ethical foundations of western communication ethics are not universally shared or approved by countries and regions worldwide. The interesting contrast to the west are Islamic states, which stretch from the

Middle East to Southeast Asia, and where Islamic tradition plays a critical role in defining ethical communication.

ISLAMIC FOUNDATIONS OF ETHICAL COMMUNICATION

Although the Judeo Christian perspective is reflected in the ethical values of the west, there is still a legal separation between church and state, which may vary from country to country, but still establishes boundaries between secular and religious law. In contrast, the Islamic world, which constitutes about one quarter of the globe's population, infuses religion into the modern state. While the commingling of religious and secular tenets in the Islamic state has been modified in a post modern era after western contact, Islam still exercises significant influence over the sociocultural lives of Islamic inhabitants (Gibb, 1964; Hourani, 1971; Hovannisian, 1985).

Tawhid, one of the most important tenets of Islam, requires obedience to the laws of Allah (Hovannisian, 1985). Human beings, according to tawhid, are obligated to know and follow Islamic codes even if sovereign law conflicts with Allah's judgment. Tawhid proclaims the supremacy of Islamic codes over sovereign law, foreign ethical systems, or personal values, and obligates the faithful to obey (Mowlana, 1989).

Within an Islamic society, all communication is evaluated as ethical based on its compatibility with the Koran. First Amendment safeguards, universal access to information, and listener choice—hallmarks of western ethical communication—are incompatible with tawhid that looks only to the Koran for ethical guidance. Similarly, a free and unfettered system of mass communication is also in conflict with Islamic tradition. All electronic communication, then, is to be judged ethical by Islamic precepts that are articulated in the Koran.

The preeminence of intellect and reason, so critical to western thought, are also incompatible with Islamic ethics (Gibb, 1964; Rahman, 1984). In fact, Hovannisian (1985) writes that intellectualism practiced apart from the Koran is "a sin against human nature—maybe even a crime" (p. 8). Clearly, western broadcasters and journalists reporting in an Islamic state are challenged by this religious imperative and may find it significantly at odds with their training and ethical tradition. Salaman Rushdie, author of *The Satanic Verses*, experienced first hand the penalty for violating tawhid, having lived in seclusion for several years until his death penalty was recently lifted by the current Iranian government but opposed by many Iranian clerics who argue that only Ayatollah Khomeini, who died in 1989, can remove Rushdie's death penalty.

A GLOBAL COMMUNICATION ETHIC?

Contrasting and seemingly incompatible ethical systems challenge the development of a global communication ethic—a standard(s) by which to judge the ethical appropriateness of messages, their availability, and the media that transmit them. Islamic and western communication ethics are more than just different: they are on a collision course. Yet, some have argued that universal values do exist in societies, and these universals may serve as a springboard for a global communication ethic. Cooper, Christians, Plude, and White (1989) suggest that some common values exist in societies which include the need to express oneself, the expectation of honesty from others, and the requirement of social responsibility from those one interacts with. In a more recent book, Christians and Traber (1997) reduce universal values to the following triad: truth telling, respect for another person's dignity, and no harm to the innocent.

Christians and Traber (1997) argue that universal values spring from the "primal sacredness of life that binds humans into a common oneness" (p. 12). Reverence for life is the foundation of human dignity and is a fundamental right of being human, according to Christians and Traber. Human dignity should be provided to all people, regardless of race, gender, age, or ethnicity, and the denial of dignity is morally repugnant.

Truth-telling is another universal value that is rooted in the sacredness of human life, according to Christians and Traber (1997). They argue that falsehood and deception are morally unacceptable because they undermine the social order of any society. If humans cannot trust others to be truthful, the social fabric of a society would crumble. In short, human beings depend on stable social systems for survival, and truth-telling is a fundamental requirement of any enduring and healthy society.

Finally, Christians and Traber identify nonviolence as a universal value that springs from the reverence of human life. They argue that Martin Luther King and Mahatma Gandhi made nonviolence more than a political strategy—it became a public philosophy etched in the moral consciousness of the human family. Nonviolence is the well from which no harm to the innocent is derived—a universal value that transcends culture and nation, according to Christians and Traber (1997).

Regardless of the values proposed for a universal communication ethic, this approach collapses when it is applied to specific cultural contexts and conflicts. Suddenly, "commonly shared" values like social responsibility and truthfulness are examined in light of multiple countries' customs and traditions, and the cultural sparks fly! Cultures can and do interpret the same values quite differently, and often do not agree on how a value ought to be expressed in a society

Truth-telling and no harm to the innocent are ideals that ring so true in theory but collide in practice with indigenous cultural beliefs. In Salaman Rushdie's case, we have an Indian-born author who writes a novel that, in his estimation, tells the truth about Islam, and he is not only sentenced to death by the Ayatollah Khomeini, but is refused entry to his homeland, India, by the Indian government. His novel, *Satanic Verses*, is considered blasphemy by many Islamic clerics, truth-telling by Rushdie, and free expression by his western supporters. Rushdie claims he is innocent of all charges: western supporters agree with him and are appalled by Khomeini's death sentence; and many Islamic clerics still consider him guilty.

Social responsibility, an important value in Islamic and western societies, has different meanings in Saudi Arabia and the U.S. From a Saudi perspective, social responsibility is synonymous with compliance to the Koran. For the U.S., this interpretation violates the very essence of social responsibility that requires volition not obedience, choice not compliance.

CONCLUSION: AN ALTERNATIVE ETHICS FOR GLOBAL COMMUNICATION

An intracultural approach is an alternative to a universal communication ethic (Shuter, 1990, 1998, in press-a and-b). An intracultural perspective acknowledges the varied and potentially conflicting ethical traditions of societies. This approach searches for the "ethical and moral constraints of a society, the deeply held cultural beliefs in communication expectations that regulate human affairs" (Shuter, in press-a, p. 8). This approach also reveals a tight relationship between a society's values, its ethical traditions, and the communication practices, policies, and laws it has developed.

An intracultural ethical perspective has the following characteristics:

(1) It is based on the assumption that ethical communication standards are wedded to culture.

(2) It relies on grounded analysis to do ethical critiques rather than universal ethical frameworks.

(3) It challenges communication ethicists to immerse themselves intellectually and emotionally in the culture of the community being studied.

(4) It requires communication ethicists to explore and explain possible effects of their personal backgrounds on their ethical analysis.

Because ethics and culture are inseparable, universal values, and the frameworks that spring from a universal approach, inevitably lead to culturally biased ethical analysis. By shifting to a grounded analytical approach, communication ethicists can carefully consider a community's indigenous beliefs, religions, and sociocultural assumptions that drive ethical perspectives about communication. Since communication ethicists are products of their own cultures, it is vital that they take stock of their sociocultural underpinnings and consider how their backgrounds affect their analysis of ethical communication. With a deep understanding of self and others, ethicists may be able to render ethical critiques of communication that reflect the complexity of culture and the subtlety and nuance of cultural value.

An intracultural perspective will also help ethicists compare the nuances of ethical traditions held by several cultures, identifying subtle differences that drive ethical clashes in communication practices and policies. Cultural comparisons made after gathering strong intracultural data can produce unique insights into communication impasses that may have once seemed insurmountable. Sensitive and yet probing intracultural analysis is a pathway to greater intercultural understanding of complex communication challenges in the 21st century and the 3rd millennium.

REFERENCES

Boafo, S. T. (1991). Communication technology and dependent development in sub-Saharan Africa. In G. Sussman & J.A. Lent (Eds.), *Transnational communication* (pp. 103–125). London: Sage Publications.

Christians, C. & Traber, M. (1997). *Communicating ethics and universal values*. Thousand Oaks, CA: Sage Publications.

Cooper, T., Christians, C., Plude, F., & White R. (Eds.). (1989). *Communication ethics and global change*. White Plains, NY: Longman.

Gibb, A.R. (1964). *Modern trends in Islam.* Chicago: University of Chicago Press.

Hourani, G. (1971). *Islamic rationalism.* New York: Oxford University Press.

Hovannisian, R.G. (1985). *Ethics in Islam.* Malibu, CA: Undera Publishers.

Hudson, H. (1990). *Communication satellites: Their development and impact*. London: Collier Macmillan.

Lent, J. (1991). The North American wall: Communication technology in the Caribbean. In G. Sussman & J. Lent (Eds.) *Transnational communication.* London: Sage Publications.

Mowlana, A. (1989). Communication, ethics, and the Islamic tradition. In T. Cooper, C. Christians, F. Plude, & R. White (Eds.) *Communication ethics and global change.* White Plains, NY: Longman.

Mueller, M., & Tan, Z. (1997). *China in the information age.* Westport, CT: Praeger.

Nilsen, T.R. (1966). *Ethics of speed communication.* Indianapolis: Bobbs-Merrill.

Oliveira, O.S. (1991). *Cultural genocide.* Sao Paulo: Edicoes Paulinas.

Rahman, F. (1984). *Introduction to Islam and modernity.* Chicago, IL: University of Chicago Press.

Shuter, R. (1990). The centrality of culture. *Southern Communication Journal, 55,* 237–249.

——. (1998). The centrality of culture revisited. In J. Martin, T. Nakayama, and L. Flores (Eds.). *Readings in cultural contexts.* Belmont, CA: Mayfield Press.

——. (in press-a). The cultures of ethical communication. In L. Samovar & R. Porter (Eds.) *Readings in intercultural communication.*

——. (in press-b). The cultures of rhetoric. In A. Gonzalez (Ed.) *Annual on International and Intercultural Communication.*

Sussman, G., & Lent, J.A. (Eds.). (1991). *Transnational communication.* London: Sage Publications.

Tefft, S. (1995, April 13). China surfs the internet, but gingerly. *Christian Science Monitor.*

Wargo, R.J. (1990). Japanese ethics: Beyond good and evil. In D. Smith (Ed.), *Philosophy east and west.* (pp. 129–138). Honolulu: University of Hawaii.

Part III
Nature and Forms of Interaction in Global Society

CHAPTER ELEVEN
Global Communication and Human Understanding

D. Ray Heisey
Kent State University

Introduction

The purpose of this chapter is to develop the argument that television, as one of the means of global communication, facilitates improved understanding between peoples of contrasting societies. It serves as a vehicle for bringing people together who have been divided by conflicting ideologies and events in history that have driven them apart emotionally and politically. Television permits viewers to see personally, albeit in a mediated fashion, the leaders of other nations which have been kept outside the circle of acceptance because of their national record of domestic and international behavior. This has the potential for breaking down stereotypical perceptions if arguments are used that are designed to identify with the viewing audience and their belief system and value structure.

Two countries that have been outside this circle of national acceptance for varying periods of time in the U.S. experience are China, the last important Communist nation, and Iran, an Islamic nation designated by the U.S. as one of the nations that supports state terrorism. The details of how these nations came to be outside the circle of U.S. close relations are well known by readers of recent history.

The new China established following the Communist Revolution of Mao Zedong in 1949 isolated itself from the free world for many years, and the U.S. rejected the legitimacy of the revolution and its subsequent government as well. This relationship continued for 23 years until President Nixon dared to go to China to make a break in Chinese-U.S. relations in 1972. Gradually over the years, with the powerful force of Deng Xiaoping's opening-up policy and economic reform begun in 1978, the relationship improved. Normal diplomatic relations were re-established and economic trade developed that have brought into existence another new China that has accepted the concept of a market economy "with Chinese characteristics." The attitudes of many people, however, and some elected representatives in Congress toward the socialist government in the People's Republic have not come along the acceptance continuum as readily as the US government's policy in recent years.

Iran, under the Shah, was a friend of the U.S. in the Middle East for many years. But that all changed when, again, a revolutionary civil war, such as occurred in China, brought down the friendly government and installed an anti-American regime that called the U.S. "the great Satan." The holding of U.S. diplomatic personnel for 444 days by this regime sealed the enmity between and the mutual rejection by these two countries of each other. Even after the release of the American hostages at the beginning of the Reagan administration, the two countries continued their mutual defiance of each other's legitimacy.

These two countries demonstrate, in their respective ways, the differences between their approaches to making a change in international perception through the use of the media and in suggesting the difference in how far along these two countries are in attempting to bridge the gap of understanding. In China's case, relations have developed to the point where the president himself was invited to come to the U.S. to engage in a summit with the U.S. president and agree to further exchanges for advancing the nature of the relationship. In Iran's case, the relationship has not yet developed that far. It was a major breakthrough, however, when Iran's newly-elected president took the bold step to address the U.S. from his own capital by means of a televised interview with an internationally recognized news correspondent, Christiane Amanpour. Both of these events will be examined in this essay as evidence of how the media can promote greater understanding between the leaders of countries at ideological conflict with each other when the human dimensions of these leaders can be observed by the other side and their words and arguments be assessed by audiences viewing them.

This chapter examines first the rhetorical event of the Chinese president visiting the United States in late October and early November in 1997 and specifically addressing the American people at Harvard University and then analyzes the interview mediated to the world by CNN of Iran's president in January 1998. The arguments and rhetorical vision of each mediated address are constructed along the lines of Kenneth Burke's concept of identification as a means of reducing the gap in understanding between opposing positions and the reported impact of each leader's attempt at identification with the audience of the other

country's people. The response of the national government in each case is
also addressed as part of the impact of these mediated events, obviously designed to influence attitudes and beliefs about the so-called opposition. A conclusion is offered regarding the potential for national leaders to use the media as a means of attempting to bring about a change in perception of certain of their world neighbors who may have been in an antagonistic relationship politically. This process of globalizing, or changing world perceptions of nations through the use of media directly available to all peoples, is now especially possible because of technology. The means for making a global impact with one's message, both verbal and nonverbal, by world leaders, can now accompany changes in national motivations and purposes.

PRESIDENT JIANG'S U.S. VISIT AND ADDRESS AT HARVARD UNIVERSITY

Although Deng Xiaoping had visited the US in 1987 on an important visit during his opening-up era, the 1997 visit by President Jiang was billed as a very significant state visit because China was now in a post-Deng era with his chosen successor in place as the acknowledged leader of the fast-reforming China. How open would he be to the pressures by the American visit on the controversial issues confronting the U.S. and China, such as human rights, Taiwan, trade, and arms sales to third-world countries? How would his American audiences receive him when there was always the possibility of protesting Chinese nationals and Chinese-Americans where President Jiang would visit? One of the wishes President Jiang had was to make a visit to Harvard University, considered by him to be the epitome of American higher education and intellectualism, certainly a forum that would be open to him for the expression of his ideas and an ideal location for his side of the dialogue to take place. The reason for choosing the Harvard address (Jiang, 1997) for closer analysis is that it was viewed as "a kind of summary of his week long state visit to the United States" and holds in it the essential arguments for "both China's new pride and peaceful intentions" (Erlanger, 1997, p. Al).

Jiang's Harvard Address

Overall, what Jiang tried to accomplish in this address was to present the Chinese view of the reformed China, no longer tied to the ideological constraints of the Mao period, but now open to the opportunities of the market economy, even though marked by the unique Chinese characteristics as developed under the creative thinking of Deng. Jiang wanted to convince his audience that China was making great progress in its reform measures and turning the corner in its own plan for domestic development and improvement, as well as getting back into the circle of the world's accepted nations in terms of international behavior.

The four central issues he discussed in the address were that (1) mutual understanding is the basis for international relations; (2) mutual exchanges are necessary so that each side can learn from the other; (3) China's path of development is unity, independence, and peace; and (4) China's opening-up and reform is comprehensive, rejuvenating, and friendly, for China is one of the nations that can be counted on to be a constructive partner in the world for peace, prosperity, and mutual development. The arguments that he used to develop these issues were drawn from his own cultural background and from his perception of the American experience. In this way he attempted to create an audience identification that would allow his viewers and listeners to see his point of view and possibly make a shift in their attitudes toward China's place in the world and in its relationship with the U.S..

In the first issue Jiang argued that much mutual understanding between China and the U.S. has taken place, but more needs to be done. He said "China needs to know the United States better and vice versa" and for this to happen, China must be approached "from a historical and cultural perspective" (Jiang, 1997, p. l). He then recounted some of the historical achievements in Chinese dynasties and the unique contributions that Chinese scientists and inventors developed. He took the opportunity to tell his American audience, "China was a world leader in science and technology for one thousand years up to the 15th century. These Chinese inventions and creations underlined the rational brilliance of coordination between humanity and nature and the integration of scientific spirit and moral ideals" (Jiang, 1997, p. 2). This point was clearly designed to help his American listeners see the Chinese perspective on mutual understanding for there was no effort to highlight American achievements that ought to be better understood by his Chinese nation.

His second issue on mutual exchanges as a way to learn from each other emphasized again how Chinese history shows that China has always been interested in learning from other peoples and cultures, constantly improving from incorporating thoughts and ideas from others. Ancient dynasties in China followed the maxim of "drawing widely upon others' strong points to improve oneself" (Jiang, 1997, p. 2). Today, he said, China is pushing forward for more extensive exchange and co-operation with the rest of the world. This point closed with an analogy from nature. "Sunlight is composed of seven colors, so our world is full of colors and splendor. Every country and every nation has its own historical and cultural traditions, its own strong points and advantages. We should respect and learn from each other and draw upon each other's strong points to offset our own deficiencies with the aim of achieving common progress" (Jiang, 1997, p. 3). This was a further attempt to emphasize the Chinese perspective which he believed Americans needed to hear.

The third issue, describing China's path of development, was intended, as well, to let Americans see his side. He explicitly said that his "observations" were to "help [America] know China better" (Jiang, 1997, p. 3). This point was developed by his describing four Chinese traditions which would assist Americans in learning to know what motivates Chinese thinking and behavior.

They are, first, the tradition of solidarity and unity by which the People's Republic of China developed among its many peoples a "new type of relationship based on equality, solidarity and mutual assistance between different nationalities" (Jiang, 1997, p. 4) and even provided regional autonomy for minority nationalities.

Second, the tradition of "maintaining independence" has been practiced for several thousands of years and after "humiliation by imperialist powers" which badly weakened China, China "has once again stood up as a giant" to find "a suitable road to development" that proceeds from its "own national conditions . . . without blindly copying other countries' models." In international relations, as well, Jiang said that China decides its "positions and policies from an independent approach" (Jiang, 1997, p. 4).

Third, the peace-loving tradition of China must be understood as part of its path of development. The ancient doctrine of "loving people and treating neighbors kindly" has been an important part of China's history for centuries, he argued. And today, more than ever, for China's commitment to modernization to be realized, peace and harmony must prevail in the world. Here Jiang emphasized the Five Principles of Peaceful Coexistence in international relations which China has subscribed to for decades, "especially the principles of mutual respect, equality and mutual benefit, and non-interference in each other's internal affairs." He promised, "We will never impose upon others the kind of sufferings we once experienced. A developing and progressive China does not pose a threat to anyone" (Jiang, 1997, p. 4).

The fourth tradition in China's path of development is "the tradition of constantly striving for self-perfection." In this section of the address, Jiang mentioned the fruits of Dr. Sun Yat-sen, "advocate of China's democratic revolution" in 1911, of Mao Zedong, liberator and builder of China's modern period, and Deng Xiaoping, leader of the "modernization drive" (Jiang, 1997, p. 5). This point brought Jiang to the final issue of the address to help America understand China better as a result of his visit and his presentation of China's thinking directly to the American people.

The fourth issue in the speech was the "reform and opening-up endeavor" that has been and is "a modern embodiment and a creative development of the Chinese spirit of constantly striving for self-perfection and renovation" (Jiang, 1997, p. 5). He laid claim in this speech to the fundamental approach that has been decided upon: "... we are right in direction, firm in conviction, steady in our steps, and gradual in our approach when carrying out reform..." (Jiang, 1997, p. 5). Chinese reform will be comprehensive, but economically it will be a socialist market economy; politically, it will be a socialist democracy; and culturally, it will be a scientific socialist culture. This reform will bring about a "rejuvenation" of the nation as it builds into "a strong and prosperous country" based upon China's own "reality and history." This attitude will promote "friendly communication" between the Chinese and American people that is the strong desire of Jiang's administration (Jiang, 1997, p. 6).

Finally, then, this reform achievement will be the basis of China's contribution to the world as China, being the largest developing country, cooperates in a friendly manner with the U.S. being the most developed country. The importance of this relationship cannot be overestimated, he claimed, for the proper "stability and development of the Asia-Pacific region and the world at large." He stressed once again in this portion of his address the importance of the basic principles of peaceful co-existence between China and the U.S. Said Jiang, "We both agree that with a view to promoting the lofty cause of world peace and development, China and the United States should strengthen cooperation and work to build a constructive strategic partnership oriented toward the 21st century" (Jiang, 1997, p. 7). Jiang's address ended with an appeal to better understanding. "I hope," he said, "that in the cause of building our own countries and promoting world peace and development, younger generations of China and the United States will understand each other better, learn from each other, enhance their friendship and strive for a better future" (Jiang, 1997, p. 8).

In addition to clarifying in the Harvard address his four central aims of China-U.S. relations—wanting to develop mutual understanding, mutual exchanges, China's path of development, and the policy of reform and opening up—Jiang in other ways attempted to demonstrate why he made this summit trip and how it fit into his overall strategy. Before he left China to go to Washington, he held a press conference in Shanghai with some Americans which disclosed his goals and strategies for the important visit to the U.S., called by Yang Dali of the University of Chicago, "the last building block for Jiang's edifice" (quoted in Wehrfritz & Liu, 1997a, p. 36). To get ready, Jiang immersed himself in American movies, videos of Deng's 1979 trip, briefing books, and information collected about "America's work ethic, pioneering spirit, and fondness for innovation-attributes he both admires and deems useful to China's modernization" (Wehrfritz & Liu, 1997a, p. 37).

Reaction to Jiang's Visit in the United States

Commenting prior to the trip on Jiang's intentions, Wehrfritz and Liu (1997a) said, "The Chinese president plans to dazzle the U.S. people by conjuring up a new, modern China that strives to put Tiananmen in the past, play a leading role in global commerce and perhaps even countenance political reform" (p. 36). These reporters also claimed that the tour Jiang designed for visiting Pearl Harbor, Colonial Williamsburg, Independence Hall in Philadelphia, the New York Stock Exchange, and Harvard University, seemed "tailored to win back America's sympathy by evoking its values" (Wehrfritz & Liu, 1997a, p. 37). His overall strategy was to sell China's agenda for the future by selling himself personally to President Clinton and to the American people. At these places he visited, Jiang played to the television cameras by swimming in Hawaii, putting on a three-cornered hat in Williamsburg, receiving the 21-gun salute and the state dinner in Washington, and ringing the opening bell on Wall Street, but also receiving two standing ovations at his Harvard address, while making "no effort to skirt the biggest and liveliest demonstrations of the entire visit" en route to the Harvard lecture hall (Wehrfritz & Liu, 1997b, p. 46).

The print media response to Jiang's visit to America was generally favorable, which certainly must have pleased the Chinese president. *Newsweek* reported that at the news conference, which Secretary of State Albright had told Jiang America would use for judging his performance, "... President Clinton and Chinese President Jiang Zemin created riveting live TV with an unprecedented public exchange on democracy and human rights." *Newsweek* further claimed that Clinton called the press conference "interesting theater" and Jiang, at the White House and throughout his visit, "proved he could take on the American press, politicians, and protesters and smile all the way" (Wehrfritz & Liu, 1997b, p.44).

The New York Times reported that Jiang would leave Washington "with his stature as a global leader enhanced," and by standing alongside the leader of the Western world, "forcefully" defending China's position on the concepts of democracy and human rights as being "relative and specific ones" ... "to be determined by the specific national situation of different countries," Jiang would benefit personally from Clinton's successful policy of "constructive engagement" (Broder, 1997, p. Al). In giving Jiang's defense of the Chinese action at Tiananmen, Broder (1997) wrote, "Mr Jiang, clearly relishing the international spotlight that the Washington summit meeting brought him, told a global television audience that a country of 1.2 billion could not progress toward economic reform and renewal without 'social and political stability'" (p. A18).

The reaction was not all positive, however, as President Jiang met protesters at many of his stops. *Newsweek* carried an entire article focusing on this theme when it published on November 3, 1997, "It's Gonna Be a Long Week: Protesters plan to give Jiang a dose of America's rambunctious democracy" (Liu & Bogert, 1997, p. 44, 45). Americans and some Chinese in America who oppose China's policies on Tibet, human rights, Taiwan, religious freedom, defending its action at Tiananmen, and other issues, took advantage of Jiang's visibility to call attention to their causes. Demonstrators, both pro- and anti- Jiang, who "numbered about 5,000" outside the Harvard's Sanders Theater, a "church-like lecture hall" (Erlanger, 1997), where Jiang spoke, helped bring national visibility to China-U.S. relations and to President Jiang personally. At his Harvard visit, Jiang gave the famous Fairbank Center for East Asian Research a newly published set of *Twenty-Four Histories with Mao Zedong's Comments,* calling it "a rich heritage of philosophy, in understanding and drawing lessons from China's history." By giving this gift and personally taking part in many other events during his visit to America, Jiang "tried to do his part," claimed Erlanger (1997), "to give post-Tiananmen China a human and unthreatening face" (p. A14).

But the exchange of ideas and spirit was mutual, as Jiang wanted it to be in his Harvard speech. For example, Jiang claimed that seeing the American demonstrations for himself, had given him "a more specific understanding of American democracy" (Wehrfritz & Liu 1997b, p.46). And Jiang had said at Williamsburg that "one of his main goals this week was to immerse himself in American tradition and culture" (Faison, 1997, p. A18). Duane Wang, a Chinese-American, said, "President Jiang will learn. He will find our system is good and he will go home with a different attitude" (quoted in Clines, 1997, p. A18).

Further evidence that Jiang may have been changed and more mutual understanding enhanced by his visit to America is that within weeks of his return to China, the "father of China's democracy movement," Wei Jingsheng, who had spent years in jail and was serving a new 14-year sentence, was released and sent into exile in the U.S. for medical treatment. Though Jiang had talked tough on human rights and defended his country's position on dissidents while in the U.S., once he returned home, he was able to make concessions that would reward Clinton for his China policy of "constructive engagement." Merle Goldman, author of *Sowing Seeds of Democracy in China,* said that "it's clear that Jiang's visit to the United States has something to do with it" (quoted in Wehrfritz, 1997 p. 42).

This view was supported by Henry A. Kissinger (1997) who said, even before the release of Wei Jingsheng, that the proper approach for dealing with human rights is not to publicly pressure the Chinese leader at a summit meeting, which "might well prove demeaning to both sides as a kind of trade in hostages. A far better test will be Chinese actions in the months ahead, when they can be undertaken as an expression of China's own decisions, not of foreign pressures" (p. 48). This, of course, is precisely what happened. Many other prisoners are still waiting for release, but *Newsweek* reported that it had learned that "Jiang was alone among China's top seven Politburo members in wanting to consider releasing dissidents, but he was evidently outvoted" (Liu & Bogert, 1997, p. 45). Prominent Tiananmen student activist Wang Dan was also later released in 1998 as a further concession on the human rights issue. In 1996, also, China had revised its laws to remove "all counterrevolutionary statutes" and John Kamm, a human-rights lobbyist, has claimed that "Beijing is considering a policy change" that could possibly set free "hundreds of convicted jailed dissidents" and "that the Ministry of Justice plans to review these cases and release prisoners whose prior activities broke no current laws" (quoted in Wehrfritz, 1997, p. 44).

On the eve of Clinton's visit to China in June 1998, Jiang further identified his goals for the second part of the Clinton-Jiang summit meetings that took place in Beijing, by sitting for an exclusive interview with *Newsweek's* Lally Weymouth. In this interview, Jiang repeated his goal of working with the United States in building "a constructive strategic partnership" and promoting "mutual understanding between our two peoples" (Jiang, 1998, p. 26). In answering the question as to what impressed him most during his visit to the U.S., Jiang said, "What impressed me most was that wherever I went, I was received in a very friendly manner by the American people. And I was given a very warm reception by the U.S. government and the various non-governmental organizations. And I was able to have very sincere and candid conversations with people from various walks of life including government officials… And these exchanges of views have proven very helpful in promoting mutual understanding between the two sides" (Jiang, 1998, p. 27).

Another indication of the softening of Chinese attitudes and of bringing about greater understanding of the issues considered important by the United States is that President Jiang agreed while he was in Washington at the summit meeting "to allow an American religious delegation to visit China, although the dates of the trip and the itinerary were not made public at the time." But in February 1998 the delegation of Rev. Don Argue, president of the National Association of Evangelicals, Archbishop Theodore E. McCarrick of Newark, and Rabbi Arthur Schneier of Manhattan, did make the trip to China to visit Tibet and talk with officials. Mike Jendrzejczyk, director of Human Rights Watch Asia, regarding this development, "suggested that the Chinese Government recognizes that it's in their interest to try to demonstrate a more open, tolerant attitude toward religious beliefs" (Shenon, 1998, p. A3).

From these statements and events, it can be seen that Jiang clearly believed that his visit to America helped him to accomplish his goals of developing a greater understanding between the Chinese and American people and their governments as well. In this sense, he used the media successfully to modify perceptions on both sides. Regardless of the protesters who received some publicity while Jiang was in the U.S., the majority of American people gave President Clinton an approval rating of 55% for his handling of foreign affairs, with 35% disapproving, and 10% with no opinion, just a couple weeks following the Jiang summit meeting which ended in early November 1997 (McAneny, 1998). And one of the leading foreign policy experts in the United States, Zbigniew Brzezinski (1998) who was National Security Advisor under President Carter, said of Clinton's foreign policy successes, "China has been handled pretty well."

PRESIDENT KHATAMI'S TELEVISION ADDRESS ON CNN

After his landslide victory of 70% of the vote in Iran in May 1997, President Khatami announced that he was interested in addressing the American people at some point, which signaled the possibility of a thaw in U.S.-Iranian relations. There was much anticipation of this address when it finally was aired on January 7, 1998 by the Cable News Network

(CNN). It was presented at 6:00 P.M. Eastern time in the U.S. and thus was in time for prime time coverage.

The address or conversation, as it was called, with CNN world correspondent Christiane Amanpour, in the office of the president in Tehran, was approximately 45 minutes in length and consisted of two major components—an opening statement and responses to a series of questions put to him by Ms. Amanpour. Analysis of the address will follow this format: elements of his statement will be presented first and then the major arguments in his answers to the questions.

Khatami's Opening Statement in the Dialogue

Khatami discussed three major issues in his opening statement. The first issue, his analysis of the American civilization, clearly was designed to appeal to the American audience and show that his attitude was intended to identify with their values and their history. He began by describing his understanding of American history as being symbolized by Plymouth Rock that portrayed the Puritans' vision of building a combination of religious freedom and democracy. He then mentioned Abraham Lincoln as symbolizing the desire of the American people to abolish slavery that had been wrong from the start. Finally, he emphasized his understanding of the American experience as a story of creating a mutual relationship between religion and liberty that the Americans found to be compatible. He ascribed American success in this endeavor as being due to the pillars of their civilization and the dialogue among civilizations on these points.

His second major issue was his analysis of the Iranian civilization. It consisted of "a blend of the Iranian heritage and the Islamic civilization." He mentioned the constitutional movement in Iran at the turn of the century that then had to struggle with a period of colonialism. The damage done by the colonial period brought about the Islamic Revolution in 1978–79 that had two directions. The first direction was its interpretation of religion that was a coupling of "religiosity and liberty." He argued that human experience has taught that the prosperous life can only be brought about through an emphasis on "religiosity, liberty, and justice." The second direction of the Islamic Revolution was "the new phase of a reconstruction of civilization." This required a dialogue between civilizations and the practice of independence.

The third main issue in his opening statement was an analysis of the interaction between the United States and Iran. This period of interaction between these two countries consisted of two important elements, in his view. The first was Iran's deep humiliation that could only be overcome by the Islamic Revolution, putting the country back on track. It was "a war of words, not weapons," he claimed, and thus should be acknowledged as a legitimate revolution. Another reason it should be so acknowledged is that it was a novel expression of religion joining with independence as two compatible pillars, just as had occurred in the American experience.

The second important element of the interaction period was that the American involvement with Iran was a story of creating "the humiliation of Iran" based on "a flawed foreign policy of domination." This domination produced three severe damages. The first was the damage to oppressed nations in many parts of the world at the hands of the United States, though specific nations were not mentioned by Khatami. The second was the dashed hopes of the peoples of the colonized world created by the U.S. The third was the damage done in the name of America which, because of its own history and experience, should have instead been standing up for freedom and independence, not oppression and domination.

In his opening statement, then, President Khatami attempted to identify with his American audience by reminding them of their history and the pillars of their civilization so that he could compare the elements of the Islamic Revolution in Iran in a favorable light with those same pillars, the unique building together of a desire for religious experience and a desire for freedom and independence. By claiming that the roots of the Iranian revolution were very similar to those of the American revolution, he wanted to place the Iranian civilization in a context with which his American viewers could more easily identify than they had been able to before. For this attempt at rhetorical identification, President Khatami received some favorable response from some Americans, which will be discussed later.

Khatami's Responses to Questions in the Dialogue

The second part of the address/conversation, consisting of responses to the questions by Ms. Amanpour, was a continuation of his rhetorical effort at putting the story of Iran in the best light possible, in terms of their own experience and perceptions. The issues brought up by Ms. Amanpour in very specific terms and answered in the question and answer period were (1) the hostage crisis, (2) the evidence that his country was a New Iran, (3) talking with the American government, not just the American people, (4) where the dialogue with America would lead (5) support of terrorism, (6) rejection of the Middle East peace process, (7) Iran's nuclear weapons program, (8) authority to pursue a genuine dialogue with others, (9) dealing with the conservatives in Iran, and (10) the future of Iran's reform movement. The essential answers to these questions given by President Khatami will now be examined.

(1) Regarding the hostage crisis, Khatami presented a long discussion, including his view that the image of Islam held in the West is erroneous because Islam invites "rationality" and "all religions to unite around God worship," it holds "the right of all human beings to determine their own destiny, and is "an enemy to no nation." Further, U.S. foreign policy is flawed because it creates "a perceived enemy" and has portrayed Islam as that enemy. He said he regrets that American feelings were hurt by the hostage crisis, but "these same feelings were also hurt when bodies of young Americans were brought back from Vietnam, yet the American people never blamed the Vietnamese people, but rather blamed their own politicians for dragging their country and its youth into the Vietnam quagmire" (p. 6). He said that his own people's feelings were also hurt by U.S. foreign policy, so the hostage crisis was "the crying out of a people against humiliations and inequalities imposed upon them by the policies of the United States," but when "there are receptive ears, there is no need for anything but discourse, debate and dialogue" (p. 6). "The events of those days," he continued, "must be viewed within the context of the revolutionary fervor and the pressures to which the Iranian nation was subjected, causing it to seek a certain way to express its anxieties and concerns" (p. 7). In sum, he said that Iran was now in a period of "stability" and was conducting its affairs "within the framework of the law," so that "everything must be analyzed within its own proper context" (p. 7).

(2) A New Iran can be seen, he said, by looking at the chanting of "slogans" as a symbol of "a desire to terminate a mode of relations which existed between Iran and the United States" (p. 7). "Our objection," he continued, "is to the type of relationship where our nation is humiliated and oppressed" (p. 8). So, with an end to this, he argued that Iran was now ready for "dialogue and understanding between two nations, especially between their scholars and thinkers" (p. 8).

(3) Talking with the American government is another matter, however, because Iran, he said, has "no need for political ties with the United States. . . . There is a bulky wall of mistrust between us and the United States' administrations" (p. 8, 9). He then went into a discussion of some of the historical events in the relationship between the two countries to support his point about the mistrust of the government.

(4) The dialogue with the American people, however, can lead to positive things, he argued, because it "should be centered around thinkers and intellectuals" (p. 9). "I believe that all doors," he continued, "should now be opened for such dialogue, and understanding and possibilities for

contact even between American and Iranian citizens should become available" (p. 10).

(5) Regarding the support of terrorism, Khatami firmly objected to the accusations, saying, "Terrorism should be condemned in all its forms and manifestations. Assassins must be condemned. Terrorism is useless, anyway, and we condemn it categorically. Those who level these charges against us are best advised to provide accurate and objective evidence, which indeed does not exist" (p. 11).

(6) On the rejection of the Middle East peace process, Khatami took a strong position against Israel, claiming that Iran has declared its opposition to the proposal because "we believe it will not succeed" and "all Palestinians" "have a right to self-determination" which should be the basis for any peace proposal. "The subject of the Middle East peace," he continued, "is one that needs a sober and pragmatic analysis. We believe that it will not succeed because it is not just and it does not address the rights of all parties in an equitable manner. We are prepared to contribute to an international effort to bring about a just and lasting peace in the Middle East" (p. 13).

(7) Khatami answered the concern about the nuclear weapons program by claiming, "We are not a nuclear power and do not intend to become one. We have accepted IAEA safeguards and our facilities are routinely inspected by that agency" (p. 13).

(8) The question of having authority to pursue a genuine dialogue with others was answered with the promise, "I'm determined to fulfill my promises, and I believe the atmosphere is conducive and would improve every day. . . . We will surely implement any policy that we formulate. It is possible that preliminary steps in certain areas might need time, but when we arrive at the policy, we will definitely carry it out" (p. 14).

(9) Regarding dealing with the conservative faction within his country that might be opposed to his opening up for dialogue, Khatami said, "Such differences of opinion are natural, and they are to be found in all societies. We should learn not to allow such differences to turn into confrontations but to direct them into their legal channels. Certainly, there are elements who are opposing our government, but so long as their opposition is practiced within the provisions of the constitution, we certainly respect them" (p. 14).

(10) Finally, on Iran's future in terms of reform when there are those there opposed to moderate reform, Khatami concluded, "Let these divisions find their meaning within their own context. Terms such as conservative, moderate and the like are more often meaningful in the West. Of course, we have differences of opinion in Iran, too. And one political tendency firmly believes in the prevalence of logic and the rule

of law, while there might be another tendency that believes it is entitled to go beyond the law. . . . I consider them natural and we need not worry about it" (pp. 15, 16).

Reaction to the Dialogue in the United States

The response of the U.S. government was favorable though cautious. President Clinton said "he would like nothing more" than to have a dialogue between the people of Iran and the people of the United States (Sciolino, 1998, p. A10) and the State Department officially insisted "that any dialogue must occur between the two governments" (Sciolino, 1998, p. Al). U.S. officials, of course, "hoped that, after nearly two decades of adamant hostility, the people-to-people contacts would eventually lead to direct talks on key issues: Iran's support for terrorism, its opposition to the Arab-Israeli peace process, and its effort to develop weapons of mass destruction" (Liu & Dickey, 1998, p. 30).

The press also was positive in its response to the Iranian president's message which was precisely what Khatami wanted to see happen in America. *Newsweek* emphasized the fact that direct governmental talks were not called for, but the positive aspects of the message were that Khatami "produced some surprises," such as his denouncing terrorism "in all its forms and manifestations" and saying that he was sorry "the feelings of the great American people have been hurt" (Liu & Dickey, 1998, p. 30).

The New York Times reported that Khatami "was at times warm and conciliatory, repeating his belief that the world had much to learn from the West, and at times harsh and condemnatory of America's past treatment of his country" (Sciolino, 1998, p. Al) and that "The attention given the interview, on a slow day in Washington, was a publicity coup for CNN, if not necessarily for Iran" (Erlanger, 1998, p. A11). In its editorial on the speech, *The New York Times* took an understandable position by saying, "Changing the tone of Iranian rhetoric about the United States is itself an achievement, and Mr. Khatami may have gone as far as he could for now, given the resistance of Iran's supreme leader, Ayatollah Ali Khamenei, to better relations with Washington... The hostility no longer serves the interest of either nation. But it cannot be eliminated or even productively addressed, if Mr. Khatami will not countenance direct discussions with the American Government" (Khatami, 1998, p. A26).

Even though some critics surmised that Khatami may have "misread American television viewers, who are mostly unaware of Iranian history and may have missed his historical references" (Sciolino, 1998, p. A10), others said that the Iranian president "particularly encouraged the

Americans when he cited Plymouth Rock as a symbol of faith and freedom," and, further quoting a senior aide, "He not only got the values right, he got the symbolism right as well, and it's so easy to misfire on that sort of thing" (Liu & Dickey, 1998, p. 30).

Other Factors

This important message by Iran's president in reaching out to the United States for an improved people-to-people, if not an eventual improvement in the political relationship, may have been the result not only of a changing political climate within Iran, as a result of the May presidential election, but of a response to some gestures that President Clinton had signaled previously. In the month before the speech by Khatami to the world audience, the Clinton Administration, much to the liking of the Iranian government, had placed "an Iranian-exile group called MKO, based in Iraq and dedicated to overthrowing the Tehran regime, on its official list of terrorist organizations," and also, after not allowing outside investors to help finance Iran's energy development, "the White House quietly agreed last summer to let a Western consortium build a natural pipeline across Iranian territory" (Depke & Bogert, 1997, p. 48).

Beyond the election that put in power the reform-minded Khatami and the gestures offered by Washington, there were other signs that signaled the changing attitude in Tehran. In December 1997, in Tehran, the 8th Organization of Islamic Conferences was held, entitled "Dignity, Dialogue & Decision," in which the Islamic nations of the world deliberated on ways to increase economic, cultural, and political cooperation with each other and throughout the region (Shakib, 1997). One of the purposes of the conference was to familiarize "the Arab people with the current way of thinking in the Islamic Iran, and creating better understanding between the Arab and Iranian peoples" (p. 35). This position was in line with Khatami's declared foreign policy "of the necessity for getting close to the Arab world" (p. 33). Furthermore, it was reported that Iran was moderating "more and more as the summit approach[ed]" in December toward the Persian Gulf states, where "the old antagonistic rhetoric which usually characterize [that of] between Iran and the six Persian Gulf Arab monarchies across the strategic waterway is absent" (p. 44). The religious leader of Iran, Ayatollah Khamenei, also showed that Islam could have cordial relations with Christians when he said, "We love Iranian Christians. They helped their Muslim fellow countrymen in the course of the Iraqi imposed war and defended their country. The Armenians and Christians are part of the

Iranian nation. I have special feelings towards Armenian fellow citizens" (Shakib, 1997, p. 43).

The official attitude toward terrorism taken by Khatami in his address on CNN was foreshadowed as well by his new Foreign Minister, Kamal Kharrazi, after his approval by the Iranian Majlis. In his press conference, Kharrazi emphasized the foreign policy of Iran as being based on "independence, equality, non-interference into internal affairs of other nations, and mutual respect." He continued, "The new government in Iran is ready to talk with other countries in order to remove any misunderstanding, to build mutual confidence, to eliminate tension and promote peace and stability in the world as a whole." He concluded, "We want expansion of political and economic relations with the countries who are capable of understanding the realities in today's Iran such as an all-out participation of the people in the recent election and the formation of the new government" (Shakib, 1997, p. 42).

In an intriguing article in *Echo of Islam,* published after the Khatami speech, Shakib (1998) wrote of Imam Khomeini's "Stance Regarding the People," in which he presented Khomeini's views again on the importance of the leader of a revolution staying close to the people by having "respect and courtesy for the people," and "giving value to the people both during and after the victory" (p. 11). Khomeini is quoted as saying, "When people don't want a public servant, he has to step down," and "Anyway, we have to go after the concept of keeping the people for as it is said, without the backing of the people, nothing can be done" (p. 11). The article argues that, based on Khomeini's philosophy, "the legitimacy of the people and the governments are the practice of Islamic commandments and only the people as a preventive force and arms can stop offences" and then quotes Khomeini again as saying, "In the Islamic Iran, the people are the ultimate decision-makers on all problems. See what the people want, give it to them. Children and adults demand freedom and independence, you should give it to them" (p. 14). If President Khatami is following the counsel of Iran's Ayatollah, one can only surmise that Khatami believes he was reading his people accurately when he called for dialogue with the American people.

In the same issue of the *Echo of Islam* (1998) as the reports above appear, President Khatami echoed his world-wide address over CNN to the United States when he is quoted as saying, "The Islamic Republic is trying to have cordial relations, based on mutual respect, with the rest of the world particularly with Muslim nations" (p. 15). It is quite clear that Iran's new president was trying to use language in the public sphere to give a changed image to the Arab nations, as well as to the West, of an

Islamic nation that can hold an independent position and at the same time offer a hand at improving cooperation through dialogue and persuasion.

The central aim of Khatami in presenting a different and changing face of his country to the United States was clearly evident in the language and arguments used in this TV conversation. The language consisted of much softer tones with conciliatory implications toward wanting to understand and to be understood in a more mutual relationship with the U.S. The harsh rhetoric was absent from his speech, but more importantly, his arguments were framed in ways to offer explanation for some of Iran's past behavior, such as the hostage crisis, and to demonstrate his attempt at understanding the history and value structure of the American people. His reference to American symbols and to what is important to Americans was used to help his audience see the Iranian perspective in some of the events that had created hostile feelings between the two countries. This positioning in the public sphere of a more moderate, reform-minded leader than had been present in the past by means of world-wide cable network news was a definitive signal to all hearers and viewers that Iran wanted to be perceived as desiring a change.

The positive but cautious response from the U.S. government, as well as leaders in the press, served Khatami's purpose well in advancing the beginnings of a new era in a previously hostile relationship. This was accomplished by means of attempting to identify with the American people, first of all, and by implication with the American government in the end, through language and argument designed for mediated acceptance. The fact that Khatami did not go as far as the U.S. government would have liked by announcing a readiness to dialogue directly must be seen simply as one of the diplomatic approaches he had available to him in light of the factions within his own country. He began at the cultural level rather than at the political level, which is a first step in trying to reach an understanding at the people-level on controversial issues.

Clinton's handling of this international event also was viewed favorably by the American people. A reserved, honest, but favorable reaction by President Clinton to the speech and the public gesture of a changing relationship with Iran—one of the last hostile nations in the U.S. sphere of influence—was welcomed by the nation. The Gallop poll measuring public opinion on how well Clinton was handling foreign affairs for America showed that 62% of the people approved of the job he was doing while only 29% disapproved and 10% held no opinion in the week of January 25 and 26, 1998, only about two weeks following the Khatami speech (McAneny, 1998). Perhaps Americans were remembering what one news correspondent reminded his viewers after Khatami's dialogue with Ms. Amanpour, when he said, "Well, they [the U.S. government] say that this is good that they're calling for this dialogue. Remember before the United States resumed its own relationship with China in the early 70s under President Nixon, there was ping-pong diplomacy, there was dialogue in effect between the people of

China and the people of the United States" (Blitzer, 1998, p. 17). This television dialogue with America may have been what Khatami intended in his own words, as "a crack in this wall of mistrust" (Khatami, 1998, p. 17).

CONCLUSION

The purpose of this chapter was to show how international leaders can use the public media to place a personal face on a national image in order to influence listeners and viewers in the "opposing" country toward a revised perception of events and characteristics that have been seen previously in an unfavorable light. By engaging in highly visible summit meetings, first in the U.S. and then in China, President Jiang wanted to convince America that his country had turned the comer and was becoming a highly respected nation in the world of law-abiding nations, nine years after Tiananmen, within a context of building positive economic reform with "Chinese characteristics." He attempted to identify with the American people by being seen in the presence of their honored and sacred cultural symbols and by giving President Clinton every advantage associated with live TV coverage at his press conference and speech at Peking University, as well as visiting and speaking directly with Chinese citizens, and finally, offering to Clinton further good-will gestures in releasing Chinese dissidents who were on Clinton's list of concerns. President Khatami, going only as far as he reasonably could be expected to politically, attempted his media coup by talking directly to America and the world, while remaining at home. He was not even able to offer governmental dialogue with the U.S., let alone any suggestion of visiting Washington or of Clinton visiting Tehran. Too much ground lay between the two countries before such talks could take place. But rhetorically, Khatami took the first step and Clinton accepted it for what it was worth. His substantial effort at laying some groundwork of mutual understanding between the two hostile countries helped to reduce the distance. The media facilitated this effort by making the channel available to him that would benefit both sides.

As both of these international leaders voiced in their rhetorical presentations to America, at the heart of the problem, as they saw it, was a need by both sides in these relationships for greater understanding of the cultural and historical contexts in which the peoples lived. Both leaders, Jiang and Khatami, acknowledging America's cultural and historical background as being responsible for where the U.S. stood on issues, wanted some recognition that the American people and government would accord the same understanding of their backgrounds. In this sense, Jiang and Khatami were asking for the spirit voiced recently by the president of the Czech Republic, Vaclav Havel, in his assessment of the conflicts between the East and the West.

Havel (1997) argues that an examination of different civilizations shows that at the core of most cultures is a basic human spirit that simply expresses itself in different ways in dealing with the concepts of legitimacy, authority, and responsibility. That spirit is one for understanding—a need inherent in all humans and societies. He says, "I see the only chance for today's civilization in a clear awareness of its multicultural character, in a radical enhancement of its inner spirit, and an effort to find the shared spiritual roots of all cultures—for they are what unites all people… The preconditions for this are genuine openness, the will to understand each other, and the ability to step beyond the confines of our own habits and prejudices. Identity is not a prison; it is an appeal for dialogue with others" (p. 300).

The call for understanding and dialogue that Havel makes is precisely the call that Jiang and Khatami made in their efforts for China and Iran to communicate with America. The interesting fact is that Havel believes the world's technology, symbolized by the media and information age, is responsible for contemporary achievements but has also been challenged recently by a contrary development described as "the recovery of an 'archetypal spirituality' that is 'the foundation of most religions and cultures'—'respect for what transcends us, whether we mean the mystery of Being or a moral order that stands above us'" (quoted in Capps, 1997, p. 312).

Capps continues his interpretation of Havel as follows: "I hear him plead that politics and politicians must cultivate new attitudes if they are to meet the challenges and opportunities of our world. He is surely correct when he observes that it is not enough for us to try to reform political methods and procedures. We must revise not our procedures but our view of reality" (p. 314). In this way, Capps quotes Havel as saying, politics becomes "the universal consultation on the reform of the affairs which render man human" (p. 315).

Capps admits that it is difficult for such a transcendent vision to be understood by Americans, but if international relationships are to be revised and reformed, and distance, whether cultural or political, is to be reduced, some sense of bringing about a clearer understanding of other cultures and values is required. Dialogue with peoples who hold different understandings and see different realities was the key rhetorical request being made by both Jiang and Khatami. Havel would say they are right in seeing this as the key. As evidenced in this case study, the public media, serving both of those nations, as well as the United States, as the receiving nation, are the means to such dialogue and human understanding.

Jiang came to America because he knew he could be seen and heard by America and Clinton went to China in order to make a difference at the human level. Khatami spoke with Ms. Amanpour, a highly-respected world correspondent, because he knew he would be seen and heard by America through this media event and possibly change the face of Iran by bringing about improved understanding.

We see in this comparative case study of China and Iran with respect to the United States that the intended changes in the perceptions of these two countries are possible because of the changing nature of the global society that is being built through the media and their pervasive and powerful effects. The community of nations cannot avoid the interdependent nature of the relationships constructed and the governments can use effectively the impact of global media on the peoples of the world for their purposes of improving human understanding.

REFERENCES

Blitzer, W. (1998). On-air comments following the transcript of President Khatami's dialogue. In M. Khatami, Iranian president favors people to people dialogue with U.S. http://www.Cnn.com/Transcripts/9801/07/wv.01htm1#top, pp. 1–17.

Broder, J. M. (1997). U.S. and China reach trade pacts but clash on rights. *The New York Times International,* October 30, pp. Al, A18.

Brzezinski, Z. (1998). Comments on CNN Crossfire, October 16. Author's notes from Crossfire.

Capps, W. H. (1997). Interpreting Vaclav Havel. *Cross Currents,* 47, No. 3, 301–316.

Clines, F. (1997). Pride and censure mix with a dash of apathy. *The New York Times International,* October 29, p. A18.

Depke, D. & Bogert, C. (1997). A powerless presidency. *Newsweek,* November, pp. 17, 48, 49.

Erlanger, S. (1997). China's president draws applause at Harvard talk. *The New York Times International,* November 2, pp. Al, A14.

——. (1998). U.S. greets Iran's overture, but insists on governmental talks. *The New York Times International,* January 8, p. A11.

Faison, S. (1997). In colonial America, Jiang dresses the part. *The New York Times International,* October 29, p. A18.

Havel, V. (1997). A sense of the transcendent. *Cross Currents,* 47, No. 3, 293–300.

Jiang, Z. (1997). Enhance mutual understanding and build stronger ties of friendship and co-operation. *The China Daily,* November 3 (pp. 1–8). http://www.ChinaDaily.com/transcript.

——. (1998). Jiang speaks out (interview with Lally Weymouth). *Newsweek,* June 29, pp. 26–27.

Khatami, M. (1998). Iranian president favors people to people dialogue with U.S. http://www.cnn.com/Transcripts/9801/07/wv.01html#top, pp. 1–17.

Kissinger, H. (1997). Outrage is not a policy. *Newsweek,* November 10, pp. 47–48.

Liu, M. & Bogert, C. (1997). It's gonna be a long week. *Newsweek,* November 3, pp. 44–45.

Liu, M. & Dickey, C. (1998). A soft signal from Iran. *Newsweek,* January 19, p. 30.

McAneny. L. C. (1998). Gallop short subjects. *The Gallop Poll Monthly* (pp. 25–59), No. 388 (January).

President Khatami addresses America. (1998). Editorial. *The New York Times International,* January 8, p. A26.

Sciolino, E. (1998). Seeking to open a door to U.S., Iranian proposes cultural ties. *The New York Times International,* Jaunary 8, pp. Al, A10.

Shenon, P. (1998). China will allow U.S. clerics to visit Tibet, group says. *The New York Times International,* January 7, p. A3.

Shakib, L. C. (1997). *Echo of Islam,* November & December, No. 161 & 162.

——. (1998). *Echo of Islam,* February & March, No. 164 & 165.

Wehrfritz, G. (1997). He's free at last. *Newsweek,* November 24, pp. 42–44.

Wehrfritz, G. & Liu, M. (1997a). Here comes Jiang. *Newsweek,* October 27, pp. 36–38.

——. (1997b). A noise in Jiang's ears. *Newsweek,* November 10, pp. 44–46.

CHAPTER TWELVE
Reconstructing Cultural Diversity in Global Relationships: Negotiating the Borderlands

Mary Jane Collier
University of Denver

The purpose of this chapter is to provide an overview of a working model of cultural diversity and global relationships that I propose to be one useful window through which to view social contact between members of diverse cultures. The model is a work in progress intended to contribute to our study of and our conduct with each other in global relationships. I invite dialogue about the claims that I advance here: that there is merit in being reflexive, recognizing our intimate involvement in the process of knowledge construction, engaging and translating multiple perspectives, focusing on relationships between macro structures/ideologies/processes and micro communicative practices, and applying what we know to who we are and what we do as we enact our intergroup relationships.

I will describe first the need to interrogate our metatheoretical/philosophical premises and positionalities and then move to clarifying my conceptualization and approach to culture and diversity through the lens of cultural identity. Then I'll present an exploratory analytical framework for coming to understand global diversity and negotiating how we relate to one another based on broader structural features including power as well as situated social praxis. Finally I'll move to propose processes of translation and dialogic negotiation as a means of enacting relationships in the borderland spaces between us.

PHILOSOPHICAL PREMISES

I bring several assumptions to the creation of such an analytical model. The first is that I am intimately engaged in the process of building knowledge and I am being/doing/becoming the role of intercultural interactant as I conduct inquiry about the process of intercultural interaction in global settings. The second is that there is benefit in reflexively examining scholar positionality. I am a middle class, middle aged, white, professional European American female who benefits each day from unearned privileges based on my race, class, and professional status, and I have experienced discrimination on the basis of my sex. I have engaged others in dialogue about relationships and power in South Africa, the Middle East, the British Isles, and the U.S. to broaden what we know in intercultural communication.

Given my research program, a third assumption I make is to recognize the value of the voices of critical theorists, post-colonialists, and feminists who call for us to interrogate positions and processes through which power and hierarchical status are achieved, and who issue strong calls for structural and situational change, and thus, emancipation. The dialogic framework for research and praxis proposed here reflects my own continuing evolution and work to incorporate such processes in a relevant manner. My overall goal is to uncover the ways in which we make sense of our diverse cultural identities and bring to light through discursive consciousness (Giddens, 1984) a clearer understanding of how we construct relationships across and within borders. I wish to build knowledge about the nature of those relationships as both/and sites and moments of time, as contestations and smooth collaborations, as based in interdependencies and dialectic differences (Baxter & Montgomery, 1996; Martin, Nakayama & Flores, 1998).

Thus, my orientation to scholarship comes from a particular place and an enduring and changing set of experiences, socialized premises, and taken-for-granted privileges. My own biases and training, experiences, and reflexive dialogue with other scholars currently orient me away from views of culture as static, homogenous groups of people with similar genetic make-up or ancestors, or for that matter, birthplaces in a particular geopolitical site. I am drawn toward views of culture as that which is socially constructed through histories, structures, and situated communicative practices by groups of people who enact their paradoxical identifications with and distinction from other groups (Collier, 1998a,b).

As intercultural communication researchers we have a great moral responsibility to examine our processes of inquiry and the consequences (intended or unintended) on the "subjects" whom we study. Through what Foucault has called regimes of truth, scholars create representations of groups and thus run the danger of reifying ideologies, structures and norms that may have profound implications on the groups (Prus, 1996). I also hold that intercultural communication scholars can contribute in various forms to the negotiation of social problematics (Deetz, 1994) through practical science (Shotter, 1993). Given that I am committed to uncovering domination and the creation of global relationships through negotiation and peacemaking rather than violence, it is most appropriate for me to continually interrogate and deconstruct overarching assumptions I bring to my research and instruction.

In terms of theoretical perspectives that I engage, my current assumptions regarding the problematic of diversity and global relationships, lead me away from objectivity, separation between being and knowing, explanation based on psychological constructs such as beliefs and attitudes (Anderson, 1996), and move me toward social constructionism (Shotter, 1993). I focus therefore more on constructs of intersubjectivity, interdependence between being and knowing, and interpretation based on reflexivity. My intent is to develop and apply an emerging and changing, and conditionally enduring, analytical paradigm that is constituted through dialogue and negotiation among scholars about the dialogue and negotiation of cultural identities and intercultural relationships in global contexts.

KEY CONSTRUCTS IN THE ANALYTICAL FRAMEWORK

Globalization and Cultural Difference

Cornel West (1993) describes a view he calls a new cultural politics of difference, that is characteristic of current discussions of globalization in an era of postmodern skepticism. The distinctive features of this view are "...to trash the monolithic and homogeneous in the name of diversity, multiplicity, and heterogeneity; to reject the abstract, general, and universal in light of the concrete, specific, and particular; and to historicize, contextualize and pluralize by highlighting the contingent, provisional, variable, tentative, shifting and changing...the diverse cultures of the majority of inhabitants on the globe smothered by international communication cartels and repressive postcolonial elites (sometimes in the name of communism, as in Ethiopia) or starved by the austere World Bank and IMF policies that subordinate them to the North (as in free-market capitalism in Chile), also locate vital areas of analysis in this new cultural terrain" (p. 204). In other words, globalization is a moment in time in which nation-state economies of the world are increasingly interdependent, new technologies have affected the nature, amount and immediacy of information available, homogenization and pluralism are concomitant, localization, and international outreach occur simultaneously, and cultural space is "shrinking and expanding" (Jenkins, 1997, p. 43).

The phrase "cultural diversity" in some countries such as the U.S. connotes more than differences in cultural backgrounds and ideologies; it also brings to mind a political standpoint calling for critique and change of the structures of domination, racist, sexist, and discriminatory policies and practices targeting members of marginalized groups, i.e., members of ethnic groups categorized as "minorities" women, gays and lesbians, elderly, and persons with physical disabilities. Global cultural diversity connotes the existence of multiple cultures existing within geopolitical boundaries (Murphy, 1996), sometimes accompanied by a critical view of power and domination.

When we discuss cultural difference, whether to advance a critique of power structures and histories of oppression, or issue a call for emancipation of different peoples, we advance and contest ideologies and premises about what is valued. In the intersections of time and space, and contact between people, ideologies are diverse, contradictory, and have their basis in different group identities, e.g., family, local community, region, ethnicity, nation-state, religion, race, sex, and political standpoint. The increased access and contact that emerge with globalization include and result in increased exposure and conflict over policies, institutions, resources, forms of contact, and hence, ideologies. Such ideologies are manifest in institutional discourse, the speeches of political spokespersons, the content of textbooks, media representations, master narratives in political speeches or print and broadcast media, forms of entertainment, as well as the everyday discourse of members of cultural groups making sense of their lives and circumstances.

Discourse in its multiple forms is the means through which cultural differences become politically engaged and diversity becomes an ideological problematic. van Dijk (1987, 1993) points out that discourse processes are the site of cognitive transformation and racism as well as reproduction of ideological forms and institutional influence within and across groups. With regard to nationalism, Duara (1993; cited in Jenkins, 1997, p. 159) contrasts what he calls the discursive meaning of nation (ideology and rhetoric) with the symbolic meaning of cultural practices such as rituals, festivals, and kinship types, noting that communicative forms constitute with whom, how, and why we align ourselves with others. Billig (1995) as well as Jenkins (1997) notes that nationalism and ideology are endemic and habitual and are reflected in discourse about what is strongly valued, moral, normal, respectable, and sinful. All such ideologies and their correlated group identities emerge in historical contexts that are often contested. Further, such ideologies emerge not only out of history and past experience but are located in present power relations, externally imposed and internally avowed boundaries, and co-constructed relationships.

The construction of ideologies, histories, and present relationships is characterized by dialectic tensions. Various spokespersons describe the paradoxical tensions and contradictions of a trend toward global unity and a simultaneous trend toward national/ethnic/regional distinctiveness. Barber (1992) names this struggle a competition between "Jihad versus McWorld" arguing that international relations and the future of humanity may be captured in the struggle between contradictory forces toward tribalism and ethnic/religious local loyalty on the one hand; and homogenized, capitalistic, mediated images and identities formed by Western corporations such as McDonalds, on the other.

Nationality, Ethnicity, Race, and Global Diversity

A discussion of global diversity must be situated within the present debate about nation-state as a viable construct and nationalism as a process of identification, as well as ethnicity as something more than ancestry. While some scholars such as Shuter (1990) argue that we should make nationality, as a site of culture, central in our research, Ono (1998) outlines several reasons to question the viability of nation-state as equivalent with or central to culture. He argues that such field dependent orientations to national groups serve to reinforce abstract stereotypes that "... assume that a people and a culture are homogeneous entities; that they function in a systemic, logically consistent fashion; that people act predictably over time; and that the most important part about communication is reading moments of idiosyncratic behavior rather than, for instance, deconstructing one's own ideological education and position and interacting in an informed and reformed way that recognizes self-consciousness about similarity and difference, generally, in daily life" (p. 198).

While nationality is a cultural designator that features geopolitical birthplace or residence, ethnicity traditionally refers to a designator based on ancestry. Jenkins (1997) outlines a social constructionist model of ethnicity based in anthropology to contrast the traditional model. The four basic premises of the model, which are also appropriately applied to the framework of diversity and global relationships being proposed here, include the following: Ethnicity is (1) based on cultural differentiation (a dialectic between similarity and difference), (2) produced and reproduced in social interaction, (3) variable and manipulable, and (4) a collective and individual, externalized and internalized identity. Jenkins (1997) notes "... although ethnicity *may* be a primary social identity, its salience, strength and manipulability are situationally contingent...No matter how apparently strong or inflexible it may be, ethnicity is *always* socially constructed..." (p. 47).

The interrelationships of race and ethnicity emerge in contested ways in scholarship about cultural diversity. Jenkins (1997) posits that ethnicity is about inclusion (us) and race is about exclusion (them); "...the former referring to group identification and the latter referring to social categorization of others" (p. 81). He argues that racial classification is a historically specific allotrope of the general ubiquitous, social phenomenon of ethnicity. Racial classifications arise in the context of situations "...in which one ethnic group dominates, or attempts to dominate, another and, in the process, categorizes them in terms of notional, immutable differences, often couched in terms of inherent inferiority and construed as rooted in different biological natures" (p. 83).

While ethnicity is not static and is socially constructed across time and space, it is also enduring to some degree. Hall (1995) asserts that a new conception of ethnicity that engages within-group as well as across-group differences may be useful in scholarly inquiry. He adds that since ethnicity is, in part, a social category that is imposed on others, such ascriptions underscore the need for attention to power and domination as factors in social categorizing. As well Ideology and social institutions and structures also must be acknowledged.

To summarize, not only are social classificatory processes and ideological forms important but also the individual enactments and reactifications of ethnic identity in combination with nationality, socioeconomic class, race, and sex. Such classifications must be viewed as intersecting and overlapping by researchers because ontologically, such group identities are experienced as interdependent. Global contact is the process through which group identities become constituted.

Cultural Identities in Global Context

In order to understand better the sites and moments of time in which members of diverse cultural groups come together and negotiate the nature of their relationships, one useful approach may be to focus on cultural identity co-construction. In this orientation, culture is an enduring and changing, situated and transcendent, co-constructed set of communicative practices and interpretive frames shared by a group (Collier, 1998a). Membership in a group is both avowed by individuals and imposed or ascribed by others, and occurs on the basis of varying degrees of voluntary and involuntary affiliation. Groups whose members share such characteristics as race, sex, ethnicity, nationality, geography, politics, organization, and social group, among others, have the potential to constitute cultural identities. Avowal and ascription occur in complex relationships and among insiders and toward outsiders. Discourse reflected in print and broadcast media as well as everyday talk include explicit references to membering and judgments about outsiders. Although as scholars we should not be too hasty in proposing insider/outsider, or center/margin dichotomies (Bhabha, 1995) neither should we ignore describing how group members do the being/becoming (Sacks, 1984) of their group identities through in-group and out-group comparison processes (Tajfel, 1978).

The window or frame through which I'm suggesting we approach the study of culture is that of socially constructed (Shotter, 1993) identity. This approach places attention upon culture as an ongoing communicative and situated accomplishment. Phillipsen (1989) describes this kind of alignment with particular groups as membering.

It is important to note that I am making a move away from defining culture as a group of people (i.e., "the Israelis") or place (i.e., "Palestine"). As well I'm moving away from the notion that the cultural identity of persons depends upon identifying with group members who come from a particular place (i.e., "Palestinians who reside in Israel") and is shaped predominantly by socialization, institutions, and ideologies common to one nation-state, Israel, where they reside. This is an implicit, but powerful premise that guides much of our cross/intercultural communication scholarship.

Within a cultural identity approach scholars recognize that cultural identities are shaped not only by the ideologies and structures of the nation-state of residence, but also by the surrounding nation-states, by broader international structures and social processes, histories, institutions (e.g., religious/spiritual and educational) and ideologies pertaining to ethnicities, social classes, and genders, as well as localized practices and community norms, not to mention individual experience and agency.

Although the geopolitical "borders" between Israel and Palestine are clearly visible and designated by checkpoints, armed military personnel, and barbed wire in some places, in others, dirt roads and contested settlement areas demonstrate that the borders "bleed" into one another (Conquergood, 1991). The boundaries between groups of people are also socially constructed and not clear to outsiders. When young women who describe themselves as Israeli, Palestinian, and Palestinian/Israeli step off the plane for a three-week peacemaking program in the U.S., it takes the U.S. American women some time to figure out who is Palestinian and who is Israeli, and for the Middle Eastern young women to understand that most but not all Latinas speak Spanish and some do not practice the

Catholic religion. Clearly coming to understand within-group conduct as well as contact between members of different groups requires more than pointing to nationality or ethnic ancestry.

Power and Agency

When proposing an analytical framework for cultural identification that will be current and relevant to global contexts, the notions of power and agency are salient and must be acknowledged. In this context of inquiry, power is a socially constructed process; the exercise of a capacity to effectively define or constitute the conditions of existence experienced by members of a group (Jenkins 1997). Power is the capacity to achieve outcomes through use of structures of domination, and the co-ordination of social systems across time and space through management of allocative and authoritative resources (Giddens, 1984). Structures of domination and resource constraints are particularly relevant to cultural diversity given that identities are socially constructed in historically and institutionally constrained sites and moments of time. Power and individual agency impact resource management based, in part, on who is defined to be insiders and/or outsiders and/or members of the margins. To illustrate, Hooks (1989) among others describes what many have called the "triple jeopardy" experienced by black, lower class, women while Hegde (1998a) describes the triple levels of dislocation experienced by migrants, contrasted with members of groups such as white European Americans with unearned advantage and privilege (McGoldrick, 1994).

McClintock (1995) reminds us that cultural identities are interpellated and articulated within and through one another in contradictory ways and in historical and structural contexts. She, among others, relates such notions of identity enactment to individual and group agency and describes a diverse set of politics of agency including coercion, complicity, dissembling, and mimicry.

Negotiating Polyvocality and Interpellation

Since the research problematic in this chapter addresses global diversity, it is important to note that cultural identities overlap and intersect. Even if a research question focuses on Afrikaners today in South Africa, understanding their communicative conduct is informed by recognizing the multitude of other groups overlapping with ethnicity and nationality, such as social class, region of residence, professional status, sex, and religion, to name a few. Those diverse group memberships bring with them socialized practices, normative prescriptions, major premises about in-group and out-groups, power, history, and the like and thus lead to individual group members speaking with diverse voices and standpoints.

There is benefit to acknowledging a multitude of voices and group identifications in addition to nationality or ethnicity. One cultural identity may emerge as salient in discourse (Collier & Thomas, 1988) as reflected in U.S. Americans interviewed by media representatives avowing their support of U.S. military action in Iraq, but interpretation of that salient identity is strengthened by knowledge of other identities, ideological, and historical forces. For instance, the responses of Saddam Hussein are best interpreted within a wider framework of his national, political, gendered, religious, and class based identities; contested histories, and, relationship with the U.S. and surrounding Arab countries, among other factors. Postcolonialists such as Bhabha (1995), Hall (1995), and Trinh (1995) describe similar ideas as antiessentialism and recognition of polyvocality. Delgado (1994) echoes such views in his description of Latino/a identity in the U.S.

Scholars such as Hegde (1998b) and Parry (1995) point out that cultural identities are often experienced more as a state of hybridity than as one or another being selected for particular contexts. In situations in which one or a hybrid identity emerges, some group members, by virtue of skin color or other external markers, have more agency and freedom to choose to feature a particular identity than other group members (Martin, 1997).

Agency and power come into play in that constraints and consequences for avowal of some identities differ. For instance, sometimes avowing racial identity alignments is viewed as advantageous to increase the numbers of persons speaking with a relatively unified voice. Blacks in South Africa explained in 1992 (Collier & Bornman, 1999) that they preferred to identify with their race rather than their separate ethnic ancestry because all blacks experienced discrimination during apartheid and blacks constitute a numerical majority in the country. Afrikaners and British preferred ethnic designators for their group identities. Thus, although I concur that race, ethnicity, and gender are social constructions, as intercultural communication researchers, we need to recognize that the character of our social relationships is based on enacting alignment with and distinctions from others, and our discourse is filled with in-group and out-group distinctions and contradictory alignments.

I suggest that we should recognize that social categories such as race and ethnicity are indeed arbitrary and changing and have been used by dominant groups to maintain resources and power. At the same time we should recognize that as humans we construct our relationships based on self-identified alignments with particular racial and ethnic groups as well as ascriptions about group identities of others (Collier, 1998b). In addition, varying degrees of freedom and choice to avow particular identities exist. (As a German, British U.S. American, I can choose when and how to reveal my ethnic background, while a dark-skinned African American does not have the same choice.) There are different consequences of having others ascribe "white" to me and "black" to an African American friend of mine.

Multiple Levels of Analysis

The tendency to oversimplify and dichotomize self and other, us and them, East and West is contrary to acknowledging interpellation, polyvocality, and the interrelationships among structures. Bernstein (1983) describes the dangers of false dichotomies and a need for us to move beyond the "Cartesian anxiety" that has characterized so much of scholarship in the humanities and social sciences. One way to do this is to incorporate macro and micro levels of analysis, and multiple perspectives to inquiry into the analytical framework.

Giddens (1984) proposes that there are three levels of phenomena through which we may study a society, or arguably cultural membering processes in international settings. These are the institutional order (ways of doing, the systematized, historical sets of organized patterns) interaction order (relationships between persons), and the individual order (what individuals think and feel) Hecht, Collier & Ribeau (1993) also posit that cultural identity may be studied from different levels or perspectives: the communal, relational, and individual. Jenkins (1997) argues further that cultural categorization can be examined along an informal to formal continuum moving from "primary socialization, routine public interaction, sexual, and communal relationships, informal group membering, marriage and kinship, market relationships, employment, administrative allocation, organized politics and official classifications" (p. 63).

Recognizing multiple levels of analysis is important in that past scholars of communication have tended to orient research programs around particular contexts of research (interpersonal relationship development or media effects, for example). Such foci place arbitrary boundaries upon what is studied, as if institutions, community norms, histories, and media do not have a role to play in intercultural relationships and intercultural relationships do not play a role in interpretation of media texts. Such contexts seem somewhat archaic when considering many of the current cultural problematics being studied in our journals. For instance Hasian & Delgado (1998) gave attention to rhetorical ambiguities and race in discourse surrounding Proposition 187 in California. They gave attention to the impact of speeches by political spokespersons, media coverage, and interpersonal everyday discourse and the role of various racial, ethnic, and nationalistic group members in constructing the discourse.

In summary, I propose that we examine cultural diversity through a combination of intersecting processes that range from historical, institutional, and ideological, to local and situated. As scholars our analytical frameworks should be consistent with the ways in which humans in general, and each of us in particular, experience and construct our worlds. Our knowledge building as scholars should be consistent with the ontological observation that humans seem to have a need for the transcendent, whether spiritual ideal or political exigency, as well as a need for situated and local recognition and a sense of place. Hegde (1998a) describes how Asian Indian immigrants create such transcendent spaces by requiring family members in the U.S. to follow traditional norms, continue family rituals, and wear traditional dress. The "idealized" nature of such images is illustrated by her example of the U.S. Asian Indians visiting India after many years away to find that the Asian Indian teenagers wear blue jeans and listen to Western music.

Our analytical frameworks need to be consistent with not only migrant experiences, but also with refugee experiences. For instance, a few months ago I conversed with a Palestinian about land and identity as we stood on the site of his father's, grandfather's, and great grandfather's former village, bombed and then uninhabited for fifty years. As we talked we looked across the hill to a new Jewish settlement. He told me what it was like to live in a refugee camp, to conduct small groups of international visitors to sites like this, to feel such a strong affinity to the spot of earth where we stood, and to see his children walk hand in hand with their grandfather, gathering herbs he pointed out, not understanding the tears in the grandfather's eyes, nor the importance of what had been lost. A little while later I visited a Jewish settlement and saw and heard about a family making a home for themselves, their children, and their people, and developing new and powerful ties to their spot of earth. These are the borderlands that are occupied, contested, the sites in which people are willing to live and die for their identities.

RECONSTRUCTIOIN THROUGH TRANSLATION AND DIALOGUE

The intersections of nationalism, racism, and ethnicism are processes in which borders between geopolitical entities and people are constituted and contested. How as scholars can we come to understand such complex processes and be open to change and be changed by engagement of such processes? I propose a reconstructionst direction that incorporates translation and dialogue as we move in/out/through the borderland spaces. Here, I add reconstruction to a previous essay in which I outlined a call for reflexivity and relevance in our intercultural communication scholarship (Collier, 1998c). McPhail (1997) presents a similar orientation in describing the utility of an approach to rhetoric that is grounded in transcendent spiritual principles of Zen and enacted through coherent dialogue that recognizes the relational implicature between self and Other. Such dialogic approaches may require some degree of translation.

Translation

Translating Across Theoretical Perspectives. There are multiple types and levels of translation as I am defining it. The first is based from an epistemological standpoint that values multiple ways of knowing (and is grounded in multiple ontological presuppositions). This kind of translation, therefore, emerges from recognizing a plurality of theoretical perspectives. The process I am calling for here begins with an appreciation that each theoretical perspective (e.g., positivist, critical, feminist, ethnographic, and reconstructionst) is developed with a set of coherent and consistent sets of assumptions about knowledge building, humans as communicating beings, as well as the being/doing/becoming constituted in communication. Translating the assumptions from one perspective to another can be beneficial. Bakhtin (1986) describes such translation as intertextuality while Deetz (1994) refers to this as dialogue across scholarly discourses.

To illustrate, it may be informative for traditional ethnographers to understand why postcolonialists critique essentialism as well as the position that there is consensus about what it takes to be a competent, accepted member of a cultural group. Postcolonialists may point out that the processes through which power emerges and is contested, and the struggle and contradictions inherent in the membering process are essential to acknowledge (Fiske, 1991). On the other hand, it may behoove postcolonial scholars to recognize the value of descriptions of situated identity enactment and coordinated human action in particular contexts.

Another example of such translation is evident in the work of Rawlins (1998). He found that juxtaposing results from social scientific studies on friendship alongside of his interpretive findings required that he clarify his own assumptions and think more about his interpretations. My own work from an interpretive perspective on interpersonal, intercultural alliances has been informed by the work of positivist approaches on uncertainty and anxiety management. For instance the discourse of Palestinian, Israeli, and Palestinian/Israeli young women being interviewed about their relationship development during a collaborative project at a peace camp included descriptions of both uncertainty about their partner and anxiety about working together (Collier, 1998b).

Translating across perspectives is not equated with methodological "triangulation" nor is this recognition of plurality the same as pluralism or relativism. As Hegde (1998b) rightly points out, it is inappropriate to argue that any and all perspectives are equally valuable or to take an "I'm OK, You're OK" standpoint. The merit of a particular perspective should be ascertained by examining the consistency in ontological and epistemological assumptions, including methodological choices for the problematic or set of questions being addressed. Then translation may provide ideas for additional constructs and contexts, or the need to interrogate taken-for-granted assumptions, about validity for example (Collier, 1998c).

This kind of translation is quite different from the most common form of advocacy and persuasive form of rhetoric in our published scholarship. This kind of translating occurs in the spirit of learning more about other perspectives and is closer to what Foss and Griffin (1995) describe as invitational rhetoric based on equality, immanent value, and self-determination. Wicke and Sprinker (1992) describe an interview with Edward Said in which he critiques the tone adopted by Cultural Studies or other scholars who appear as badgering and authoritative, and his horror at dogmatic orthodoxies "paraded as the last word." They also note that he calls for a more critical, engaged, perhaps dialogical approach in which "alternatives are present as real forces" (p. 241).

Sometimes such translation requires moving from one language game to another (Wittgenstein, 1974) and understanding that constructs mean different things from different theoretical orientations. For example, from a social psychological view of the role of media in adaptation, individuals have freedom in their choices of action, thus agency is individually based and behavior may be measured through asking individuals to respond to a questionnaire or answer questions about their individual conduct in an interview. From a critical theory perspective, hegemonic, frequent, and positive media portrayals of members of the dominant group, reinforcement of general master narratives, myths of equality and individual freedom in advertisements, all serve to constrain individual agency and provide only an illusion of choice. I am suggesting, therefore, that in order to be an informed evaluator of global research conducted by intercultural researchers, one must be familiar with a range of theoretical perspectives, be able to judge each on the basis of its own coherence and consistency among assumptions and conclusions, as well as be able to translate across perspectives.

Translating Traditional Contexts to Global Problematics. In order to make sense of diverse global relationships we need to reach beyond what were traditionally called intercultural, interpersonal, international, public, or mediated contexts of research and practice. Historically, the "channel" used was the indicator of the type of communication being studied; in other words, face-to-face contact between individuals or group representatives was called intercultural communication, and international communication was the term used for messages transmitted across national borders through mediated or technologically formed channels. Today, international communication is more often described as contact taking various forms in which global, transnational, national, ethnic, and regional identities are constituted and contested (Friedman, 1994). More specifically, international communication scholars spend time investigating messages in multiple forms, within multiple structures, and across diverse contexts, and a critique of power is often seen. For instance, topics that have emerged include globalization as an economically normative force and newer critiques of development programs creating dependency and a bias toward Western capitalism (Keith, 1997).

Translating Knowledge and Praxis. A call for dialogic translation of this last kind refers to an ongoing, collaborative process that bridges epistemological and ontological premises with praxis. Another way of talking about this form of translation is to move from analysis or "talk about" to "talking with" others. I argued earlier that all of us are engaged participants in the process of being/doing/becoming our group identities and global relationships. On that basis I recommend scholars inform our knowing by engaging in dialogue and "doing the being" (Sacks, 1984) of relationships that take the form of alliances. I'm thinking here of dialogue that can take the form of allies rather than adversaries, through which identities, relationships, and positions on issues are negotiated in the midst of contradictions, conflict, and struggle, as well as collaboration. Scholars with other scholars, researchers and respondents, respondents and their broader communities, etc., can form alliances. Allies do not necessarily agree; within this research program, allies might share the goal of understanding intercultural contact and developing knowledge about how they are implicated with each other in creating and overcoming structures of domination. Cornel West (1993) describes a similar dialogic relationship as "...forging solid and reliable alliances of people of colour and White progressives guided by a moral and political vision of greater democracy and individual freedom in communities, states, and transnational enterprises—i.e., corporations, and information and communications conglomerates" (p. 217).

Dialogue

Dialogue is the process through which relationships are formed, challenged, maintained, and changed. Buber (1972) is often described as one of the primary philosophers of dialogue. He defines what he terms "...genuine dialogue—no matter whether spoken or silent—where each of the participants really has in mind the other or others in their present and particular being and turns to them with the intention of establishing a living mutual relation between himself [sic] and them" (p. 19). The Public Dialogue Consortium, based in New Mexico, is an example of a newly formed informal organization of communication teachers and students whose purpose is to "facilitate the institutionalization of better forms of communication than generally exist in public and about public issues" and "devotes itself to create places for nonadversarial conflict resolution and improve communication among citizens" (Pearce & Littlejohn, 1997, pp. 198–199).

The members of the Public Dialogue consortium base part of their approach on The Public Conversations Project that was started in the U.S. in 1990, in which dialogue is contrasted with debate. Dialogue is based on the sharing of personal experiences and listening to the experiences of others, and includes spontaneous sharing and is interactive, and the goal is to uncover important differences and similarities among all present. The goal of dialogue about public issues in this setting is to understand. Dialogic conversation values "... listening more than speaking, understanding more than explaining, and respect more than persuasion" (Pearce & Littlejohn, 1997, p. 214).

Thinking reflexively about the underlying assumptions and who benefits and who risks most with such a model of dialogue, several concerns may arise. Postcolonialists and critical theorists, for instance, may be tempted to critique the dialogic approach described above as obviously created by scholar/practitioners who take for granted their unearned privilege and hierarchical status. Critics could point out that those who call for dialogue seem to presume that we all have equal agency and freedom of choice, and seem to ignore the standpoint that histories of racism, sexism, and imperialism constrain our actions and relationships with one another, and institutional policies and ideologies affect all of us differently. They might argue that members of groups who have experienced discrimination and oppression, and who have been co-opted into silence or beaten into submission, have the most to lose in such a conversation.

After reflecting on such concerns, I propose a kind of dialogue that starts with academic engagement of ideas including postmodern skepticism and deconstruction of taken-for-granted assumptions and master narratives. It is based in examination of diverse voices and alternative perspectives through extended conversations and intercultural relationships with other academicians. For example, Halualani (1998), in an auto-ethnographic/performative/critical essay, describes her experience growing up as a "mixed Japanese/Hawaiian/English woman in a patriarchal monoculture" (p. 265). In this essay her goal is to "narrate and perform my struggle with the hopes that uniform, homogenized, American Culture can be both critiqued and contested and that all of us, on our own, can form/inform/re-form who we are, how we are to act and live, and what we are to believe" (p. 265).

The kind of dialogue I am describing here listens and speaks, and for those of us like me in positions of unearned privilege, includes hearing the experiences of those with less privilege and more privilege, about what we take for granted and how often and how much we gain from our status. It is based on the kind of reflexivity that includes reflecting with those who are more similar to us, as well as talking with those who are critical of systems from which we benefit.

Dialogue is a process that requires some amount of structure, as I've discovered in work with Palestinian, Israeli, Palestinian/Israeli, and U.S. American young women who work on building alliance relationships during a three-week peacemaking program (Collier, 1998b). The structuring, in the way of particular communication exercises and skills of listening and clarifying, seems to work best when collaboratively constructed and critiqued and deconstructed with staff and participant feedback, and then reconstructed throughout the process of relationship development.

Another example to illustrate my claim that we must engage in dialogue in order to move forward or beyond protracted conflict is my experience facilitating a preconference session at the 1998 International Communication Association conference in Jerusalem. I invited representatives of various peacemaking youth organizations from diverse Arab and Israeli political/religious/social perspectives to teach members of the academic community about their intercultural experiences.

I learned about the need to continually de/reconstruct the structures as well as the processes and interpretations of the content of our messages.

For instance, the site of the session in Jerusalem became immediately contested because this location required Palestinians who wanted to attend the session to be granted permission and obtain special visas from the Israeli government months in advance, and such permissions are not easily obtained for many Palestinians who have been in prison for any length of time (quite a large number since the actions of 1948). As well, I learned that my original ideas for discussion topics and format (e.g., having staff and participant representatives from the youth organizations share their goals, philosophies, and exemplar experiences) carried much more risk and held less promise for learning new or practical information for the Palestinians. I learned of the differences in orientations by developing relationships with some of the staff members and investing the time to arrange visits to their offices in East Jerusalem or the West Bank.

I continue to be committed to the need for confronting bias, interrogating taken-for-granted privileges and assumptions, and I am also committed to finding what Bhabha (1995) calls "third spaces" and Anzaldua (1987) describes in her poetry and prose as the "borderlands." As one who still believes in peacemaking through dialogue, my current belief is that we are better served by dialogue that allows for recognition of co-opting, counterhegemony, diverse forms of agency, silencing, confronting, and collaborating. In order to recognize the complexity and contradictions inherent in such dialogue, we can embrace the contradictions as dialectic tensions, to recognize the alternatives not as either/or choices but as both/and dimensions. Recognizing the dialectical tensions that are profoundly evident all around us is a start toward moving beyond the philosophical lens of dualism that currently limits our knowledge about and construction of relationships with others (Bernstein, 1983).

In order to reconstruct the study of global diversity and relationships, dialectic tensions can replace the overly simplistic notions of either—or, us—them, center—margins. In an interview with Wicke & Sprinker (1992) Said discusses the Palestinian views of nationalism, calling it both essential to survival and the enemy, both positive and problematic. He noted that for oppressed peoples living in an occupied territory there is no substitute for nationalism that can lead to liberation, thus he affirms as well as rejects the idea of nativist subjectivity.

Cross-addressing: Resistance Literature and Cultural Borders (described in a book review by Harlow, 1997) also captures the nature of the contradictory tensions and paradoxical identity issues of hybridity, alterity, and marginality that are relevant across international settings today. Conflicting ideological directives and narratives of binarisms and universalisms are highlighted, and essays argue against such easy dichotomies as black/white, male/female, traditional/ modern material/ spiritual, and/or, self/other. "Border Criss-crossing" is made pan-American, according to Harlow, in the essay by Manuel Martin-Rodriquez about Ron Aria's "The Road to Tamazunchal." Lien Chao's "The Collective Self" examines Sky Lee's "Disappearing Moon Café" and the constructions of a Chinese Canadian community.

CONCLUSION: CONTINUING THE CONVERSATION

Dialogue can contribute to decolonialization by providing concrete counter examples to open up spaces for understanding and emancipation. Such spaces need not be sites of domination, but sites in which ideas are contested and assumptions challenged. We can embrace the complexity of culture as social construction that has been added to the previously limited conception of culture as territorially contained place, as Alvarez (1995), an anthropologist who studies the U.S. Mexico border, argues. He holds that the borderlands graphically illustrate the conflicts and contradictions that occur in a hierarchically organized world, and how ideologies and individuals clash and challenge our disciplinary perspectives on what constitutes culture, social harmony, and equilibrium.

Another example that illustrates the relevance of the borderlands and the need for dialogue that negotiates diversity, is a review by Murphy (1996) of the work of Pat Mora, a Chicana writer. Mora writes about "nepantla" a Nahuatl word meaning "place in the middle." Mora is described as being in the middle of her life, having been raised in the Tex-Mex borderlands, and having a psychic and cultural identity as Mexican American. Her writings capture the pain and advantage of negotiating both sides, recognizing biases, and developing views of multiple spaces, and arriving at the position that nurturing variety is central, not marginal, to democracy.

In summary, the model of dialogue that I am proposing here is based on my own experience and privileged observation of the potential of intercultural relationships in managing social problematics in the borderlands. Relationships in which multiple cultural identities are being enacted require an intention to maintain and enhance the relationship, an investment over time, a commitment to negotiation, and an openness to reflexive and critical engagement. The model I propose will evolve and change through collaboration and dialogue with others holding different perspectives and standpoints. It is my hope and my intention to participate in extending our abilities as scholars/practitioners/participants to reconstruct and build relationships that move us toward greater peacemaking throughout the world.

NOTE

* My use of first-person voice is consistent with my recognition that who I am and what I have been taught influences what I write and claim to know.

REFERENCES

Alvarez, R. R. (1995). The Mexican-U.S. border: The making of an anthropology of borderlands. *Annual Review of Anthropology, 24*, 447–470.

Anderson, J. A. (1996). *Communication theory: Epistemological foundations*. New York: The Guilford Press.

Anzaldua, G. (1987). *Borderlands/La Frontera: The new mestiza.* San Francisco: Spinsters/Aunt Lute.

Bakhtin, M.M. (1986). *Speech genres and other late essays*. (C. Emerson & M. Holquist, Eds., V. McGee Trans.) Austin: Univ. of Texas Press.

Barber, B.R. (1992, March). Jihad Vs. McWorld. *The Atlantic Monthly*.

Baxter, L. & Montgomery, B. (1996). *Relating: Dialogues and Dialectics*. New York: Guilford Press.

Bernstein, R. J. (1983). Beyond objectivism and relativism: Science, hermeneutics, and praxis. Philadelphia: Univ. of Pennsylvania Press.

Bhabha, H. K. (1995). Signs taken for wonders. In B. Ashcroft, G. Griffiths, & H. Tiffin (Eds.), *The Post-Colonial Studies Reader* (pp. 29–35). New York: Routledge.

Billig, M. (1995). *Banal nationalism*. London: Sage.

Buber, M. (1972). *Between man and man*. New York: Macmillan.

Collier, M. J. (1998a). Researching cultural identity: Reconciling interpretive and post-colonial perspectives. In D. V. Tanno & A. Gonzalez (Eds.) *Communication and identity across cultures (International and Intercultural Communication Annual, Vol. XXI)* (pp. 122–147). Thousand Oaks, CA: Sage.

——. (1998b, July). *De/reconstructing the boundaries: Dialectic tensions in intercultural interpersonal alliances among Middle*

Eastern young women. Paper presented at the International Communication Association conference, Jerusalem, Israel.

——. (1998c, February). *Reconstructing our Identities as Intercultural Scholar/Practitioners: Radical Ideas about Reflexivity and Relevance*. Paper presented at the Western States Communication Association conference, Denver, Colorado.

Collier, M. J. & Bornman, E. (1999). Core symbols in South African intercultural friendships. *International Journal of Intercultural Relations, 23*, 133-156.

Collier, M. J., & Thomas, M. (1988). Identity in intercultural communication: an interpretive perspective. In Y. Kim & W. Gudykunst (Eds.), *Theories of intercultural communication.* (pp. 99–120). International and Intercultural Communication Annual, XII. Thousand Oaks, CA: Sage.

Conquergood, D. (1991). Rethinking ethnography: Towards a critical cultural politics. *Communication Monographs, 58,* 179–194.

Deetz, S. (1994). The future of the discipline: The challenges, the research and the social contribution. In S. Deetz (Ed.) *Communication yearbook 17*, (pp. 115–147). Thousand Oaks, CA: Sage.

Delgado, F. P. (1994). The complexity of Mexican American identity: A reply to Hecht, Sedano and Ribeau and Mirande and Tanno. *International Journal of Intercultural Relations, 18,* 77–84.

Duara, P. (1993). De-constructing the Chinese nation. *Australian Journal of Chinese Affairs, 30*, 1–26.

Fiske, J. (1991). Writing ethnographies: Contribution to a dialogue. *The Quarterly Journal of Speech, 77*, 330–335.

Foss, S. A. & Griffin, C. L. (1995). Beyond persuasion: A proposal for an invitational rhetoric. *Communication Monographs, 62*, 2–18.

Friedman, J. (1994). *Cultural identity and global process.* Thousand Oaks, CA: Sage.

Giddens, A. (1984). *The constitution of society*. Berkeley: University of California Press.

Hall, S. (1995). New ethnicities. The myth of authenticity. In B. Ashcroft, G. Griffiths, & H. Tiffin (Eds.), *The Post-Colonial Studies Reader* (pp. 223–227). New York: Routledge

Halualani, R. T. (1998). Seeing through the screen: A struggle of "culture." In J. Martin, T. Nakayama, & L. Flores (Eds.) *Readings in cultural contexts*. (pp. 264–274). Mountain View, CA: Mayfield.

Harlow, B. (1997). Cross-Addressing: Resistance literature and cultural borders. (Book review). *Research in African Literatures,* 12–22.

Hasian, M. & Delgado, F. (1998). The trials and tribulations of racialized critical rhetorical theory: Understanding the rhetorical ambiguities of Proposition 187. *Communication Theory, 8*, 245–270.

Hecht, M. L., Collier, M. J., & Ribeau, S. (1993). Ethnic identity. In *African American communication.* Thousand Oaks, CA: Sage.

Hegde, R. (1998a). Translated enactments: The relational configurations of the Asian Indian immigrant experience. In J. Martin, T. Nakayama and L. Flores (Eds.) *Readings in cultural contexts* (pp. 315–321). Mountain View, CA: Mayfield.

——. (1998b). View from elsewhere: Locating difference and the politics of representation from a transnational feminist perspective. *Communication Theory,*

Hooks, B. (1989). *Talking back: Thinking feminist, thinking Black.* Boston: South End.

Jenkins, R. (1997). *Rethinking ethnicity, arguments and explorations.* Newbury Park, CA: Sage.

Keith, N. W. (1997). *Reframing international development: Globalism, postmodernity and difference*. Thousand Oaks, CA: Sage.

Martin, J. (1997). Understanding whiteness in the United States. In L. Samovar & R. Porter (Eds.) *Intercultural communication: A reader* (pp. 54–62). Belmont, CA: Wadsworth.

Martin, J. N., Nakayama, T. K. & Flores, L. A. (1998). A dialectical approach to intercultural communication. In J. Martin, T. Nakayama, & L. Flores (Eds.) *Readings in cultural contexts*. (pp. 5–14). Mountain View, CA: Mayfield.

McClintock, A. (1995). *Imperial leather*. New York: Routledge.

McGoldrick, M. (1994). Culture, class, race, and gender. *Human Systems: The Journal of Systemic Consultation & Management, 5*, 131–153.

McPhail, M (1997). *Zen and the Art of Rhetoric*. New York: SUNY Press.

Murphy, P. D. (1996). Conserving natural and cultural diversity: the prose and poetry of Pat Mora. *MELUS* (March). 1–8.

Ono, K. (1998). Problematizing "nation" in intercultural communication research. *Communication and Identity Across Cultures* (International and Intercultural Communication Annual) 21, D. V. Tanno & A. Gonzalez (Eds.) (pp. 193–202). Newbury Park, CA: Sage.

Parry, B. (1995). Problems in current theories of colonial discourse. In B. Ashcroft, G. Griffiths, & H. Tiffin (Eds.), *The Post-Colonial Studies Reader* (pp. 36–44). New York: Routledge.

Pearce, W B., & Littlejohn, S. (1997). Moral conflict. Newbury Park, CA: Sage.

Phillipsen, G. (1989). Speech and the communal function in four cultures. In S. Ting-Toomey & F. Korzenny (Eds.) *Language, communication, and culture: Current directions (International and Intercultural Communication Annual) XII*, (79–92). Newbury Park, CA: Sage.

Prus, R. (1996). *Symbolic interaction and ethnographic research.* Albany: State University of New York Press.

Rawlins, W. (1998). Writing about *Friendship Matters*: A case study in dialectical and dialogical inquiry. In B. Montgomery & L. Baxter (Eds.) *Dialectical approaches to studying personal relationships* (pp. 63–82) Mahwah, NJ: Lawrence Erlbaum.

Sacks, H. (1984). On doing 'being ordinary." In J. M. Atkinson & J. Heritage (Eds.), *Structures of social action: Studies in conversation analysis* (pp. 413–429). Cambridge, UK: Cambridge University Press.

Shome, R. (1996). Postcolonial interventions in the rhetorical canon: An "other" view. *Communication Theory, 6*, 40–59.

Shotter, J. (1993). *Conversational realities: Constructing life through language*. London: Sage.

Shuter, R. M. (1990). The centrality of culture. *Southern Communication Journal, 55*, 237–249.

Tajfel, H. (1978). Interindividual and intergroup behaviour. In H. Tajfel (Ed.), *Differentiation between social groups* (pp. 27–60). London: Academic Press.

Trinh, T. M. (1995). No master territories. In B. Ashcroft, G. Griffiths, 7 H. Tiffin (Eds.), *The post-colonial studies reader* (pp. 215–218). New York, NY: Routledge.

van Dijk, T. (1987). *Communicating racism: Ethnic prejudice in thought and talk.* Thousand Oaks, CA: Sage.

——. (1993). *Discourse and elite racism.* London: Routledge.

Wicke, J. & Sprinker, M. (1992). Interview with Edward Said. In M. Sprinker (Ed.), *Edward Said: A critical reader* (pp. 221–264). Cambridge, MA: Blackwell.

West, C. (1993). The new cultural politics of difference. In S. During (Ed.) *The cultural studies reader* (pp. 203–217). New York: Routledge.

Wittgenstein, L. (1974). *Philosophical grammar*. Oxford: Blackwell.

CHAPTER THIRTEEN
Social Exchange in Global Space

Sue Wildermuth
University of Minnesota-Minneapolis

Computer-mediated communication (CMC) is "synchronous or asynchronous electronic mail and computer conferencing, by which senders encode, [primarily] in text, messages that are relayed from senders' computers to receivers'" (Walther, 1992a, p. 52). The 1990's have seen a drastic increase in the variety and volume of CMC. For many "CMC is no longer a novelty, but a communication channel through which much of [their] business and social interaction takes place" (Walther & Burgoon, 1992, p. 51). Although the United States and Western Europe have seen the most dramatic increases in on-line communication, people from all over the world use the Internet. Due to this global participation in on-line interaction, a new type of interpersonal relationship is becoming increasingly common. This type of relationship can be labeled "computer-mediated" and refers to an interpersonal relationship that is formed and maintained primarily through computer-mediated interaction. Thanks to the Internet, more and more people are forming international friendships. But, because international visits, phone calls, and letters are often expensive, time-consuming, and impractical, these friendships are primarily maintained only through on-line interaction. Researchers who have been studying this new form of interpersonal relationship have reached varied and contradictory conclusions about how such relationships form and function and about the impact that these relationships have on cultural diversity.

The purpose of this chapter is to expand our knowledge of computer-mediated interaction by focusing on how interpersonal relationships are formed and maintained globally over electronic mail (e-mail). The literature of CMC interpersonal relationship research is reviewed, and empirical evidence is presented to examine the formation and maintenance of interpersonal relationships in a global space. Implications to human communication in a global society are also discussed.

COMPUTER-MEDIATED COMMUNICATION

Computer-mediated communication includes a wide variety of options, but e-mail is by far one of the most popular (Garton & Wellman, 1995; Sproull & Kiesler, 1986; Walther, 1996). The average person on-line conducts more business (both personal and professional) by e-mail than by any other form of CMC (Berleant & Liu, 1995). As Negroponte (1995) points out, "The single biggest application of networks is e-mail. ... [I]n the next millennium e-mail ... will be the dominant interpersonal telecommunications medium, approaching if not overshadowing voice within the next fifteen years" (p. 183, p. 191). Electronic mail is a CMC option that "allows the computer user to send and receive a 'letter' addressed to a particular individual, almost instantaneously, by typing information into his/her computer" (McCormick & McCormick, 1992, p. 380). Garton & Wellman (1995) stress that as a form of communication, e-mail has the following five characteristics: (1) it is asynchronous, (2) it allows rapid transmission and reply, (3) it is text-based, (4) it is primarily dyadic, and (5) it allows message storage and text manipulation.

Although the study of CMC is a rapidly growing field, a great deal of past research has been primarily technical or descriptive in nature, focusing almost exclusively on the use of CMC in organizational settings (Garton & Wellman, 1995; Golden, Beauclair, & Sussman, 1992). However, as CMC becomes more global and the social nature of CMC becomes more prevalent, CMC research has started to shift focus. Currently, many researchers study the social and cultural implications of CMC and examine how individuals use CMC options such as e-mail to initiate and maintain socioemotional relationships. This relational literature shows that three conflicting views dominate current thought on how CMC influences relationships. First, CMC can be seen as highly impersonal with little potential for relationship formation. Secondly, CMC can be construed as supportive of interpersonal interaction but more restrictive than face-to-face interaction so that personal relationships take longer to form and develop. Finally, CMC can be viewed as so personal that individuals self-disclose rapidly and form intensely intimate relationships on-line.

Closely tied to each of these interpersonal relationship perspectives is a cultural perspective. In relation to cultural diversity, one can either view the advent of on-line interpersonal relationships as a phenomenon that will negatively impact cultural diversity, a phenomenon that will have little to no effect on cultural diversity, or a phenomenon that will increase or enrich cultural diversity. The following section discusses each of these perspectives in detail.

Impersonal CMC

Early studies into relational CMC assert that all relationships on-line are task-focused and impersonal (Parks & Floyd, 1996; Schaefermeyer & Sewell, 1988). For example, Picot, Klingensburg, & Kranzle (1982) found that users considered CMC as most appropriate for tasks that required little personal interaction. Rice and Love (1987) compared CMC with face-to-face interaction and described CMC "as less friendly, emotional, or personal and more serious, businesslike, and task orientated [and thus, not] as appropriate for getting to know someone" (pp. 88-89). Three perspectives have been developed to explain the impersonal nature of CMC interaction: Reduced Social Context Cues Theory, Media Richness Theory, and Social Presence Theory. All three theories state that computer-mediated communication is inherently impersonal due to the restricted nature of the medium. On CMC social cues are weak and thus, the lack of indications of job title, status level, race, age, gender, or appearance, restricts interpersonal relationship formation (Sproull & Kiesler, 1986). In addition, CMC is a lean media and has limited ability to provide immediate feedback or to support various types of verbal and nonverbal cues (Daft & Lengel, 1986). Thus, the computer is a medium most appropriate for conveying information that is simple or task-focused. Finally, the limited channels available within CMC results in less attention paid by the user to the presence of other social participants. As social presence declines, messages are more impersonal (Walther, 1992a).

Connected to this view of on-line relationships is the media prediction that an increase in interpersonal relationships on-line will negatively impact cultural diversity. Stoll (1995) and Postman (1992) have predicted that widespread use of the Internet for relationship formation will lead to cultural fragmentation and decay. They see the Internet as a medium that, by facilitating anonymity, eliminates diversity. This perspective stresses that individuals who interact on-line are unable to share the unique aspects of their cultures due to the constraints of the on-line context discussed in the above paragraph. Because there are no visual, nonverbal, or social cues, cultural aspects such as dress, skin color, and culturally-specific non-verbals are not readily available. In addition to these constraints, the primary on-line language is English, so if one wants to interact on the Internet, he or she often must do so in English rather than in their native language. Thus, according to this perspective, the very nature of the on-line environment is expected to eliminate linguistic and cultural diversity.

Interpersonal CMC

In contrast to the impersonal view, more recent work on CMC stresses that interaction on-line is often socially-oriented. For example, a survey of COMSERVE, CRTNET, and PSYCHNET users conducted by Schaefermeyer & Sewell (1988) found that 63.8% of respondent's e-mail activity was social in nature. McCormick and McCormick (1992) found that more than half of the messages sent by undergraduates on a university's e-mail system were socioemotional. In addition, much of the international communication that is taking place on-line occurs for personal rather than professional reasons.

Walther (1992a) developed a theory called the Social Information Processing Perspective to explain how interpersonal relationship development occurs on-line. The theory is based on the assumption that communicators in CMC are driven to develop relationships (Walther, 1996). In order to achieve this goal, newly acquainted users form impressions of their communication partners through the text-based information they receive. Based on these original impressions, participants test their assumptions about the other through repeated applications of knowledge-gaining strategies. Over time, the information gained through these strategies accumulates until the users feel that they "know" the other, and thus, a relationship is formed (Walther, 1996). This framework acknowledges that there is less socioemotional information conveyed per CMC message because of the lack of nonverbal cues. However, it also recognizes that over time participants are able to adjust to that deficiency by developing strategies to overcome the textual nature of CMC (Walther, 1996). In this way, interpersonal relationships are possible, even in a cues-limited context such as the Internet.

This interpersonal view of CMC relates to the media prediction that an increase in interpersonal relationships on-line will not significantly impact cultural diversity. Researchers supporting this cultural prediction claim that culture is an inherent part of an individual's personality and that therefore, cultural diversity will continue to be a factor in on-line relationships (Walther, 1996). While these researchers recognize that cultural cues such as language or skin color may not be as readily available in an on-line context as they are in a face-to-face context, they stress that cultural differences still exist in individuals. These differences may just take a little longer to be acknowledged or expressed due to the restraints of the on-line environment.

Hyperpersonal CMC

The most recent research on CMC reveals neither a lack of personal interaction nor a slow rate of interpersonal communication. Instead, research findings reveal that interpersonal relationships on-line often rapidly surpass the level of self-disclosure, intimacy, and attachment found in parallel face-to-face interactions (Garton & Wellman, 1995; McCormick & McCormick, 1992). Walther (1996), labeled this aspect of CMC, "hyperpersonal communication." Although no formal theory addresses the development of hyperpersonal communication on-line, Walther (1996) proposed a combination of theoretical communication concepts (i.e., receiver, sender, channel, and feedback) to begin explaining the phenomenon.

Walther's explanation can be summarized into four main points. First, CMC receivers magnify the perceptions that they form about their communication partners (Walther, 1996). Because there are so few social context cues available on-line, the cues that do exist take on added significance and are used extensively when forming one's perceptions of the other. Secondly, CMC creates a context in which the communication sender has extensive control over self-presentation. As Walther (1992b) states, "[w]ith more time for message construction and less stress of ongoing interaction, users may ... take the opportunity for objective self-awareness, reflection, selection and transmission of preferable cues" (p. 229). Third, communicating over the computer can be an asynchronous channel which, in addition to allowing senders time to edit their messages before mailing them, also gives the sender the freedom to interact when he or she chooses. Finally, Walther (1996) proposes that CMC feedback provides a behavioral confirmation loop. In CMC interactions the channels available for additional information are limited and it is more difficult for us to receive information that is contradictory to our expectations. Thus, on-line we are more likely to continue to behave as if our perceptions are the reality, which in turn, influences that reality to become similar to our perception (Walther, 1996). As all of these factors interact with one another, they create a context for extremely personal and emotional interaction.

The hyperpersonal view leads to the media prediction that an increase in interpersonal relationships on-line will actually have a positive impact on cultural diversity. Researchers such as Gronbeck (1990), Gurak (1997), and Jones (1995) claim that the Internet has the power to expand cultural diversity by increasing the likelihood that people will be exposed to multiple cultures and by providing a new context for cultural formation. Thus, these researchers believe that the increased exposure to individuals of various backgrounds, cultures, religions, and so on, that is available on-line allows people to expand their cultural awareness. Moreover, researchers examining on-line environments such as chat rooms, MUDS, MOOS, and list-serves claim that the Internet actually increases cultural diversity by allowing people new contexts within which to create alternative cultures and societies (December, 1995).

SOCIAL EXCHANGE THEORY

With such wide disparity in CMC relational research and theory, how do we know what approach to take toward understanding the creation of relationships over a global medium such as e-mail? In addition, when thinking about on-line interaction in a global society, how can we reconcile the various cultural predictions that are currently tied to each of the three existing relational CMC theories? The answer stems from a closer examination of the assumptions underlying those theories. Each of the three relational CMC theories is based on the general assumption that people engage in communication to fulfill goals (whether that goal be to accomplish a task, to form a friend, or to find a lover), and that they choose to interact over CMC when they see the Internet as best suited to helping them achieve those goals. This assumption is at the core of a communication theory called social exchange.

Numerous researchers have hinted that a better understanding of relationship formation over CMC could be gained by viewing CMC through the perspective of social exchange. For example, Walther (1996) stated that an impulse to communicate in ways we find *rewarding* is inherently human and easily enacted via technology. Rice and Love (1987) stated that the new medium of CMC removed some of the traditional *costs* of communication (roles, norms, structure, procedures, politics, etc.) while creating others, and Parks & Floyd (1996) stated that CMC raises new *opportunities and risks* for the way individuals relate to one another. However, no formal application of social exchange theory to the CMC context has yet been conducted. Therefore, this study empirically examines interpersonal relationship formation over e-mail through the lens of social exchange. The results of this empirical research are then discussed in relation to their impact on interpersonal CMC research, and in relation to their impact on predictions about how CMC influences cultural diversity.

This study will focus on self-disclosure as the small part of interpersonal relationship formation behavior to be examined. Self-disclosure plays a decisive role in relationship formation (Berger & Bradac, 1982), and leads to such positive relationship benefits as increased social support, improved relationships, and higher levels of trust and intimacy (Altman & Taylor, 1973). Self-disclosure occurs when the communication is of personal information which is genuine, and which is not readily available from some other source. A true act of self-disclosure is voluntary and the specific content of the act concerns the self (Rosenfeld, 1979). Altman and Taylor (1973) describe self-disclosure along two dimensions: breadth and depth. Breath refers to the number of topics and the different ways in which the topic is approached, while depth examines the personal or intimate nature of the topic being discussed.

Social Exchange

While the term social exchange theory actually encompasses multiple theories of exchange, these theories share basic assumptions about what factors influence human behavior. Nye (1978) summarizes the generative assumptions underlying social exchange theory as follows: (1) Humans seek rewards and avoid costs. Rewarding behavior is defined as behavior that leads to pleasures, satisfactions, gratification, and positive emotion, while costly behaviors lead to sorrow, disappointment, guilt, anger, fear, or other negative emotions; (2) Costs being equal, humans will choose the alternative that is expected to supply the most rewards; (3) Rewards being equal, humans will choose the alternative which is expected to exact the fewest costs; (4) Immediate outcomes being perceived equal, humans will choose the alternative that is expected to provide the best long-term rewards; and (5) Long-term outcomes being perceived as equal, humans will choose the alternative which is expected to provide the best immediate rewards.

In addition to sharing basic assumptions about what factors motivate human behavior, social exchange theories also share assumptions about how individuals evaluate costs and rewards. One primary factor affecting an individual's perception of cost and reward values is social context (Roloff, 1981; Taylor, Altman, & Sorrentino, 1969). Altman and Taylor (1973) stress that two main context factors determine whether a situation will accelerate or inhibit relationship formation. These factors are situational formality and situational confinement. "The more formal the situation, the less rapid the social penetration process" (p. 162). In formal situations, the rules are clearly defined. There are specific social norms regulating what a person should or should not do. Certain behaviors and customs are appropriate, social roles are rigid, and role status is evident. For these reasons, formal situations usually involve high risks. This perception of high costs restricts self-disclosure. On the other hand, informal situations are much more relaxed, social roles are not quite so dominant, and perceptions of cost are lower. Thus, these situations are more conducive to self-disclosure.

According to Altman & Taylor (1973), situational confinement refers to "the general ease with which persons can leave an interpersonal

..ationship, particularly during its formative stages" (p. 163). In the majority of contexts, the more confining the situation, the higher the perception of potential cost and the slower the social penetration process. Conversely, in a situation with little confinement, the participants know that they can easily escape the interaction if they so choose. This knowledge should lead to perceptions of low potential cost and thus, higher rates of self-disclosure.

This study examines both the impact that a situational variable (i.e., interacting in an e-mail environment) has on individuals' evaluations of costs and rewards, and the impact that those cost and reward evaluations have on actual behavior (i.e., self-disclosure levels). As in any communication situation, there are multiple cost and reward evaluations involved in self-disclosure over e-mail. However, this study only examines the one most unique to CMC interaction: perceptions of personal anonymity.

Personal Anonymity

Because e-mail users are unable to see the person with whom they are interacting, e-mail helps to reduce self-awareness and may make people believe they are operating more anonymously (Eisenberg, 1994; McCormick & McCormick, 1992; Weisband & Reinig, 1995). In addition to this perceived anonymity, there are also e-mail options such as anonymous mailers which actually allow the original source of a message to be concealed (McMurdo, 1995). Finally, the use of aliases on e-mail is common and also allows one to remain anonymous (Myers, 1987), as e-mail users do not need to give out their real name if they do not want to. On e-mail, an individual can be whomever he/she wants to be, can say whatever he/she wants to say, and can do whatever he/she wants to do. There is no fear of ruining a personal reputation or harming a "real-life" image because individuals on-line are hidden from censorious eyes (Scharlott & Christ, 1995).

Another type of anonymity is concealed appearance. On e-mail it does not matter whether you are old or young, tall or short, unattractive or a supermodel. Because there are no face to face interactions, there are no limitations based on appearance. States one user, "All messages have an equal chance because they all look alike." Thus, the on-line environment is often seen as more utopian, and less concerned with status based on physical beauty.

Perceptions of anonymity are also facilitated by the very mechanics of interacting over e-mail. The act of typing in a password before accessing one's e-mail account conveys the perception to the user that messages are not accessible by anyone but the sender (Weisband & Reinig, 1995). In addition, hitting the delete key implies that messages are erased and increases perceptions that e-mail messages are private and easy to eliminate. Finally, reduced social cues and few reminders of a message's intended audience reinforce the illusion that e-mail communication is private and thus, that one's identity and self-disclosures are also private (Weisband & Reinig, 1995).

Moreover, research also shows that electronic mail serves to reduce the status barriers normally present in face to face interactions (Garton & Wellman, 1995; Kiesler, Siegel, & McGuire, 1984; McMurdo, 1995). On-line, people are less able to convey status, power, and prestige cues contextually or dynamically. Because "e-mail reduces status cues, status-induced imbalances are also weaker. E-mail can encourage more open and equal discussion" (Garton & Wellman, 1995, pp. 440–441).

Although e-mail facilitates *perceptions* of individual anonymity and privacy, the truth is that "e-mail is a notoriously leaky way to communicate. Users tap out their messages and send them wending through a patchwork of electronic conduits, only to discover on occasion that the wrong person has received the message" (Eisenberg, 1994, p. 128). In addition, deleted messages are often automatically saved in back-up files and can be resurrected by someone who was not the intended recipient. This fact is especially significant when we realize that we can be held legally accountable for e-mail messages we send and receive. E-mail appears anonymous and private on the surface; however, the more one knows about privacy laws and computer technology, the more one realizes that e-mail is neither. Numerous reports of on-line fraud have been reported. Because participants in on-line relationships are often at the mercy of the honesty of the other individual, there is a high occurrence of on-line betrayal. Thus, depending upon e-mail as one's only source for information about the other participant is risky and involves high levels of trust that the other's self-disclosure is accurate.

As discussed earlier, self-disclosure is a behavior that is potentially rewarding, leading to such benefits as increased social support, improved self-esteem, and deeper interpersonal ties. However, self-disclosure is also a behavior that is potentially costly, leading to such detriments as humiliation, fear, punishment, and relationship dissolution (Altman & Taylor, 1973; Sunnafrank, 1990). Based on the situation (i.e., formal or informal, confining or non-confining), self-disclosure can be perceived as either rewarding or costly. Applying social exchange theory, it is logical to assume that when the costs of self-disclosing are low enough that the rewards outweigh them, amounts of self-disclosure should increase. Likewise, when the costs of self-disclosing are so high that they outweigh the rewards, self-disclosure should decrease (Altman & Taylor, 1973).

Because many users see the e-mail context as informal, they tend to believe that individuals on-line do not feel constrained by social rules or limited by "the approval of society." They accept the rationale that it is psychologically easier for people to self-disclose when they feel no one is looking. Thus, they believe that self-consciousness decreases on-line because the fear of negative repercussions is reduced. "Possessed of the mistaken but reassuring notion that paperless communication is not really communication at all, e-mail users shed the self-consciousness characteristic of their lives ... and accelerate disclosures" (Eisenberg, 1994, p. 128). In addition, these users tend to believe that anonymity on e-mail gives individuals a large amount of control over the evolution of a relationship. Users can respond to another's message when they want to (the same day or three days later), they can respond for as long as they want to (a two-line message or a three-page one), and finally, they can respond *if* they want to (Berleant & Liu, 1995). Thus, the e-mail context is not seen as involving a high level of confinement. Individuals who interact on e-mail feel as if they can cut off communication at any time without negative consequences (Golden, Beauclair, & Sussman, 1992; Kim & Raja, 1991). All an individual has to do is type a few simple commands to change their on-line pseudonym, and no one can contact the "person" they used to be. These users see e-mail as a low-cost context, a situation with low formality and low confinement.

However, for individuals who understand the risks involved in on-line communication, the e-mail situation becomes both confining and formal. These users know that they can be quickly traced and thus, cannot escape an e-mail interaction with ease. In addition, they are aware that they must be very, very careful of what they say on-line because they do not really "know" who might be in their "unintended" audience, or whether or not the person they are interacting with is honest. These users see e-mail as a high-cost context, a situation with high formality and high confinement.

Thus, depending upon their perspective, e-mail users may see the on-line context as facilitating anonymity and thus, low in costs (i.e., informal, and non-confining), or as inhibiting anonymity and thus, high in costs (i.e., formal and confining). Given the related possible cost and reward evaluations, the following three hypotheses are posed:

H1: A negative relationship should exist between perceived anonymity of e-mail and perceived costs associated with e-mail use.

H2: A negative relationship should exist between cost perceptions for self-disclosure on-line and breadth of self-disclosure on-line.

H3: A negative relationship should exist between cost perceptions of self-disclosure on-line and depth of self-disclosure on-line.

Hypothesis 1 states that low levels of anonymity lead to high amounts of perceived on-line costs. Hypotheses 2 and 3 state that high amounts of perceived on-line costs lead to low amounts of on-line self-disclosure. The final hypotheses connect these ideas by directly linking levels of anonymity with amounts of on-line self-disclosure.

H4: A positive relationship should exist between the perceived anonymity of e-mail and breadth of on-line self-disclosure.

H5: A positive relationship should exist between the perceived anonymity of e-mail and depth of on-line self-disclosure.

METHODS

Participants

The participants in this study were current college students enrolled in Speech-Communication classes at a large Midwestern university. The final subject population consisted of 32 people (16 pairs). Participants were female, primarily white and came from mainly middle class backgrounds. Female subjects were chosen due to sampling convenience and to eliminate error variance caused from mixed-gender communication.

Procedure and Instruments

After signing consent forms for participating in an e-mail interaction, each participant filled out a questionnaire that contained items addressing the participant's general demographics and her level of familiarity with e-mail. They then were asked to complete a questionnaire that consisted of 20, 9-point questions (with 1 representing strongly disagree, 9 representing strongly agree, and 5 representing neutral). These Likert-scale questions addressed whether or not individuals perceived the e-mail context as facilitating or inhibiting anonymity.

For the next three weeks the students were asked to communicate over e-mail with their relationship partner for a minimum of one hour per week (approximately 10 minutes per day). At the end of the project, participants were asked to come to a final debriefing session. At this session, participants filled out a questionnaire accessing their perceptions of the costs and rewards involved in their self-disclosure on-line. The questionnaire contained 35 Likert-scale items addressing the costs and rewards that participants felt were involved in the interaction, and the perceptions the participants had of the breadth, depth, and reciprocity of their self-disclosures. In addition, two open-ended questions allowed the participants to relay their own personal feelings and reactions to their on-line experiences. Finally, the participants were asked to sign a release form before turning in the hard copy transcripts of their interactions.

Data Analysis

The researcher and two trained assistants coded the transcripts from this study. Coding involved looking at the number of total messages, the number of different topics discussed, and the intimacy level of topics discussed. Coders rated the intimacy of self-disclosures using a 9-point Likert-type scale (Chaikin, Derlega, Bayma, & Shaw, 1975). The first step to data analysis involved selecting the self-disclosure units on the interaction transcripts. For this study, self-disclosure units were defined as independent clauses or sentences in which the subject described herself in some way, told (or implied) something about herself, or referred to some affect she experienced. The two coders each coded copies of 10% of the transcripts, then they compared statements, discussed oppositional views when they occurred, and reached a consensus. One coder then finished the coding. A reliability check was done again at the end of the statement coding to check for coder drift, with a second coder once again coding a final 10% of the transcripts. An aggregate reliability check was done on the shared 20% of coding to compare the number of units or segments of material obtained by each of the coders. Aggregate reliability is expressed as a percentage difference between the sum of the number of units obtained by each coder. A coefficient of zero represents perfect accuracy. The aggregate reliability score of the two coders was .03, suggesting that the two coders perceived similar numbers of units in the data.

The second step to coding was to rate each of the statements for intimacy level and code for topic. Using a 9-point Likert-type scale (Chaikin, Derlega, Bayma, & Shaw, 1975), each statement was given a number between one and nine; one being very low intimacy, nine being extremely intimate. Two coders coded the first 10% of the transcripts for intimacy with a coding reliability alpha of .9064. Then one coder finished the coding. A verification check was done at the end with a final 10% of the data being coded by the second coder with a coding reliability alpha of .91. Finally, as a last verification check on coding, the coded intimacy ratings for subjects' self-disclosures were correlated with the subject's perceptions of how intimate their self-disclosures were. The Pearson Product-Moment Correlation Coefficient was significant, $r(30) = .55$, $p < .001$.

Next, each statement was assigned a topic code based on Altman and Taylor's (1966) list of intimacy-scaled self-disclosure items. The statements were divided into 13 categories: love and sex (significant other relationships and sexual experiences); family (husband, kids, pets); parental family (brothers, sisters, parents); hobbies (music, books, movies, etc.); physical appearance; money/property; relationships (friends, teachers, coworkers); attitudes/values; opinions; emotions/ feeling; school/work; biographical; and other. Cohen's (1965) kappa, or *k*, is a coefficient of interjudge agreement for nominal scales that demonstrates the proportion of agreement between coders, after chance agreement is excluded. A *k*-value of above .70 represents satisfactory interjudge reliability for this study. Two coders coded the first 10% of the transcripts for topic level with a *k*-value of .89. Then one coder finished the coding. A verification check was done at the end with a final 10% of the data being coded by the second coder with a *k*-value of .91.

Finally, Chronbach's alphas were computed for all scales and nested design ANCOVAs were used to test hypotheses in which the two covariates were amount of pair-wise interaction (case) and past e-mail experience. In this study the reliability between items addressing anonymity was strong with a standardized item Cronbach's alpha of .87. Questions addressing the perceived costs of e-mail interaction had a reliability of .80. In addition, reliability was .81 for questions addressing perceived amounts of on-line self-disclosure (breadth and depth), and .77 for questions addressing how much familiarity participants had with e-mail.

RESULTS

The results of ANCOVA, using case (pair-wise interaction) and past e-mail experience as covariates, show that Hypothesis 1 was not confirmed [$F(1,1,30) = 2.0$, $p = .18$]. That is, no significant correlation was found between total perceived anonymity of the e-mail context and total perceived costs of interacting over e-mail. ANCOVA shows significant correlation [$F(1,1,30) = 3.84$, $p = .05$] between a participant-reported measure of the total costs perceived related to e-mail interaction and a coded measure of self-disclosure breadth (total number of different topics discussed) for Hypothesis 2. However, the strength of the relationships between the dependent variable and the covariates were weak, as assessed by a partial eta square, with case accounting for .07% and past experience accounting for .09% of the dependent variable. Among those subjects that reported perceiving high costs involved in e-mail interaction, the mean number of total topics discussed was 8.65. Among those subjects that reported perceiving low costs involved in e-mail interaction, the mean number of total topics discussed was 10.98.

Thus, a negative correlation exists between total perceived costs of e-mail and coded self-disclosure breadth. ANCOVA also shows that no correlation exists [F (1,1,30) = .60, p = .82] between a participant-reported measure of the total costs perceived related to e-mail interaction and a coded measure of self-disclosure depth (overall intimacy level of self-disclosures) in Hypothesis 3.

To summarize, support was found for the hypothesis that high perceived on-line costs for interaction lead to low amounts of self-disclosure breadth on-line. However, no correlation was found between perceived on-line costs and amounts of self-disclosure depth.

The results of the ANCOVA used to test Hypothesis 4 show that there is a significant correlation [F (1,1,30) = 28.1, p = .00] between a participant-reported measure of how much e-mail facilitates anonymity and a coded measure of self-disclosure breadth (total number of different topics discussed). However, the strength of the relationships between the dependent variable and the covariates were weak, as assessed by a partial eta square, with case accounting for .11% and past experience accounting for .13% of the dependent variable. Among those subjects in the low anonymity group, the mean number of total topics discussed was 8.375. Among those subjects in the high anonymity group, the mean number of total topics discussed was 9.9375. Thus, a positive correlation was found between perceived anonymity and coded self-disclosure breadth.

A significant correlation [F (1,1,30) = 6.05, p = .00] was also found for Hypothesis 5 that specifies the relationship between a participant-reported measure of how much e-mail facilitates anonymity and a coded measure of self-disclosure depth (i.e., overall intimacy level of self-disclosures). However, the strength of the relationships between the dependent variable and the covariates were weak, as assessed by a partial eta square, with case accounting for .15% and past experience accounting for .17% of the dependent variable. Among the subjects in the low anonymity group, the mean intimacy level of self-disclosures was 5.43. Among the subjects in the high anonymity group, the mean intimacy level of self-disclosures was 7.87. Thus, a positive correlation was found between perceived anonymity and coded self-disclosure depth.

DISCUSSION AND IMPLICATIONS

This chapter's aim was to accomplish two goals. The first goal was to reconcile the three prominent relational CMC theories with one another by integrating them with concepts from social exchange theories. Empirical evidence for such integration was obtained by examining the relationship between perceptions of the CMC context and amounts of interpersonal communication on-line. It was hypothesized that perceptions of the CMC context would influence perceptions of the costs and rewards involved with interacting on-line, which in turn, would determine the breadth and depth of an individual's on-line self-disclosure. The connection between perceptions of the CMC context and an individual's evaluation of the costs and rewards involved in on-line communication was not supported in this study. However, support was found that cost and reward perceptions determine breadth of self-disclosure (consistent with past research on self-disclosure and social exchange), and that perceptions of the e-mail context as either facilitating or inhibiting anonymity determine the breadth and depth of on-line self-disclosure. Thus, it can be concluded that the on-line context is not inherently restrictive or supportive of interpersonal relationship formation. Rather it is an individual's perception of that context which determines his/her behavior. This result provides a new and appropriate tool for viewing Internet relationships and helps explain a broader spectrum of behaviors in CMC interaction than research on interpersonal CMC has been able to do in the past.

The second goal of this study was to examine the controversy existing between media predictions indicating that intercultural uniqueness would be negatively impacted by the expansion of on-line interaction and media predictions indicating that cultural diversity would be positively affected by the expansion of on-line interaction. The results of this study can illuminate the cultural diversity controversy in three ways.

First, in contrast to Stoll's (1995) and Postman's (1992) prediction that widespread use of the Internet for relationship formation will lead to cultural fragmentation and decay, the present study found a positive correlation between perceptions of Internet anonymity and levels of on-line self-disclosure. In other words, the anonymous nature of e-mail led to more open and honest disclosure about a wide variety of topics, including topics such as religion, culture, beliefs, and opinions. Thus, the results of this study would indicate that the anonymous nature of the on-line environment does not limit cultural diversity, but rather, facilitates diversity by increasing self-disclosure between the on-line participants.

Secondly, the results of this study can also be used to help illuminate the reasoning behind the prediction that an increase in interpersonal relationships on-line will not significantly impact cultural diversity. As stated earlier, this perspective believes that the restraints of the on-line environment will require that participants take a little longer to become aware of cultural differences, but that those differences will still exist and will not increase or decrease. This study provides an explanation for this prediction by demonstrating that perceptions of the on-line context are primary determinants of on-line behavior. Thus, individuals who view the on-line environment as a mere extension of an off-line environment will be likely to perceive cultural interaction on the Internet as similar to cultural interaction anywhere else. Therefore, these individuals will interact with others in much the same way on-line as they do off-line and continue to enact the same behaviors and search for the same types of information in relation to cultural differences as they did before. Based on the results of this study, it is easy to see how some people may remain the same on-line or off-line in regard to their cultural awareness and cultural uniqueness.

Finally, the results of this study support the prediction that on-line interactions will positively impact cultural diversity by increasing the likelihood that people will be exposed to multiple cultures, and by providing new contexts for cultural formation (Gronbeck, 1990; Gurak, 1997; Jones, 1995). This study implies three plausible explanations for the above prediction. First, an individual may perceive the on-line context as a place safe to express cultural uniqueness that he or she is not available to express off-line due to social constraints. Because the Internet is seen as low in formality and situational confinement, users may feel less restrained to express themselves and their culture on-line. In this case, a positive impact on cultural diversity would be found because people are able to share their cultural background without fear of negative repercussions. Secondly, individuals may perceive the on-line context as an international and intercultural forum that is an ideal place to meet people from different cultural and religious backgrounds. This perspective of the Internet as a global utopia is extremely popular in current media. Individuals who accept this viewpoint may seek out the Internet as a place to increase their cultural awareness. In this case, an Internet user would be actively seeking new experiences and new intercultural relationships and they would therefore encourage on-line diversity. Finally, individuals may perceive the Internet as a forum not only for learning more about other cultures, but also for getting actively involved in creating new ones. An individual with this perspective would work hard to establish MUDS, MOOS, chat rooms, or other unique societies that could only be found on the net. Over time, these interpersonal organizations or groups would develop new, unique, and powerful cultures of their own. In this way, cultural diversity would be increased as entirely new cultures are formed and matured.

In conclusion, as the Internet supports world-wide interaction, research into the cultural background of on-line participants is vital for understanding the dynamics of human communication and relationship development in the upcoming global society. In other words, computer-mediated communication and relationships have become so globally based, it is critical for researchers to further examine the issue by taking culture into account in order to provide more reliable information for global citizens to better understand each other and live more productive lives.

REFERENCES

Altman, I., & Taylor, D. A. (1973). *Social Penetration: The Development of Interpersonal Relationships*. New York: Irvington.

Berger, C. R., & Bradac, J. J. (1982). *Language and Social Knowledge: Uncertainty in Interpersonal Relations*. London: Edward Arnold.

Berleant, D., & Liu, B. (1995). Robert's Rules of Order for e-mail meetings. *Computer, 28,* 84–85.

Chaikin, A. L., Derlega, V. J., Bayma, B., & Shaw, J. (1975). Neuroticism and Disclosure Reciprocity. *Journal of Consulting and Clinical Psychology, 42*, 13–19.

Cohen, J. (1965). Some statistical issues in psychological research. In B.B. Wolman (Ed.), *Handbook of clinical psychology* (pp. 95–121). New York: McGraw-Hill.

Daft, R. L. & Lengel, R. H. (1986). Organizational Information Requirements, Media Richness and Structural Design. *Management Science, 32*, 554–571.

December, J. (1995). Transitions in studying computer-mediated communication. [On-line]. Available: http://www.december.com/cmc/mag/jan/1995/html.

——. (1996). Units of Analysis for Internet Communication. *Journal of Communication, 46,* 14–38.

Eisenberg, A. (1994). E-Mail and the New Epistolary Age. *Scientific American, 270*, 128.

Finn, J. (1996). Computer-Based Self-Help Groups: On-Line Recovery for Addictions. *Computers in Human Services, 13*, 21–39.

Garton, L., & Wellman, B. (1995). Social Impacts of Electronic Mail in Organizations: A review of the Research Literature. *Communication Yearbook 18,* 434–453.

Golden, P. A., Beauclair, R., & Sussman, L. (1992). Factors Affecting Electronic Mail Use. *Computers in Human Behavior, 8*, 297–311.

Gronbeck, B.E. (1990). Communication technology, consciousness, and culture: Supplementing FM-2030's view of trans-humanity. In M. Medhurst, A. Gonzalez., & T.R. Peterson (Eds), *Communication and the culture of technology.* (pp. 3–18). Pullman, WA: Washington University Press.

Gurak, L.J. (1997). Utopian visions in cyberspace. [On-Line]. available: http://www.december.com/cmc/mag/1997/feb/gurak/html.

Jones, S.G. (1995). Understanding community in the information age. In S.G. Jones (Ed.), *Cybersociety: Computer-mediated communication and community*. (pp. 10–23). Thousand Oaks, CA: Sage.

Kiesler, S., Siegel, J., & McGuire, T. W. (1984). Social Psychological Aspects of Computer-Mediated Communication. *American Psychologist,39*, 1123–1134.

Kim, M., & Raja, N. (1991, May). *Verbal Aggression and Self-Disclosure on Computer Bulletin Boards.* Paper presented at the Annual Meeting of the International Communication Association. Chicago, Illinois.

McCormick, N. B., & McCormick, J.W. (1992). Computer Friends and Foes: Content of Undergraduates' Electronic Mail. *Computers in Human Behavior, 8,* 379–405.

McMurdo, G. (1995). Netiquettes for networkers. *Journal of Information Science,21,* 305–318.

Myers, D. (1987). "Anonymity is Part of the Magic": Individual Manipulation of Computer-Mediated Communication Contexts. *Qualitative Sociology, 10,* 251–266.

Negroponte, N. (1995). *Being Digital.* New York: Alfred A. Knopf.

Nye, F. I. (1978). Is Choice and Exchange Theory the Key? *Journal of Marriage and the Family,* 219–232.

Parks, M. R., & Floyd, K. (1996). Making Friends in Cyberspace. *Journal of Communication, 46,* 80–97.

Picot, A., Klingensburg, H., & Kranzle, H. (1982). Office technology: a report on attitudes and channel selection from field studies in Germany. *Communication Yearbook 6*, 149–160.

Postman, N. (1992). *Technopoly: The surrender of culture to technology.* New York, NY: Vintage Books.

Rice, R., & Love, G. (1987). Electronic Emotion: Scoioemotional Content in a Computer-Mediated Communication Network. *Communication Research,14,*85–108.

Roloff, M. E. (1981). *Interpersonal Communication: The Social Exchange Approach.* New York: Sage.

Rosenfeld, L. B. (1979). Self-Disclosure Avoidance: Why I Am Afraid to Tell You Who I Am. *Communication Monograph,46,* 63–74.

Rosenfeld, L. B., & Kendrick, W. L. (1984). Choosing to be open: An empirical investigation of subjective reasons for self-disclosing. *Western Journal of Speech Communication, 48*, 326–343.

Schaefermeyer, M. J., & Sewell, E. H. (1988). Communicating by Electronic Mail. *American Behavioral Scientist, 32*, 112–123.

Scharlott, B. W., & Christ, W. G. (1995). Overcoming Relationship-Initiation Barriers: The Impact of a Computer-dating System on Sex Role, Shyness, and Appearance Inhibitions. *Computers in Human Behavior, 11*, 191–204.

Sproull, L., & Kiesler, S. (1986). Reducing Social Context Cues: Electronic Mail in Organizational Communication. *Management Science, 32,* 1492–1512.

Stoll, C. (1995). *Silicon snake oil.* New York: Dell Publishing.

Sunnafrank, M. (1990). Predicted Outcome Value and Uncertainty Reduction Theories: A Test of Competing Principles. *Human Communication Research, 17*, 76–103.

Taylor, D. A., Altman, I., & Sorrentino, R. (1969). Interpersonal Exchange as a Function of Rewards and Costs and Situational Factors: Expectancy Confirmation-Disconfirmation. *Journal of Experimental Social Psychology, 5*, 324–339.

Walther, J. B. (1992a). Interpersonal Effects in Computer-Mediated Interaction: A Relational Perspective. *Communication Research, 19*, 52–90.

——. (1992b). A Longitudinal experiment on relational tone in computer-mediated and face to face interaction. *Proceedings of the Hawaii International Conference on System Sciences, 4*, 220–231.

——. (1996). Computer-Mediated Communication: Impersonal, Interpersonal, and Hyperpersonal Interaction. *Communication Research, 23*, 3–43.

Walther, J. B., & Burgoon, J. K. (1992). Relational Communication in Computer-Mediated Interaction. *Human Communication research, 19*, 50–88.

Weisband, S. P., & Reinig, B. A. (1995). Managing User Perceptions of E-mail Privacy. *Communications of the ACM, 38*, 40–47.

CHAPTER FOURTEEN
Conflict, Globalization, and Communication

Leda Cooks
University of Massachusetts, Amherst

Writing about conflict, globalization, and communication seems a daunting task, especially since those who theorize about globalization seem unclear as to whether the concept refers to events happening simultaneously around the globe (like audiences watching/reading about Bill Clinton and Monica Lewinsky) or some underlying global pattern or structure which accounts for or explains all conflict processes. Theorizing about globalization and communication has taken the form of macrostructural analysis of information flow and structures or institutions that either support or restrict such flow, or alternatively has been theorized as microlevel instances of conversation where self/other distinctions are drawn and somehow implicate larger patterns of identity distinctions in a global society. Admittedly, the former is more likely than the latter, although several recent studies in cyberspace and other diasporic public spaces have attempted to make connections between culture, community, and conversation in a global space (e.g., Hawes, 1998; LeBesco, 1998; Mitra, 1996; Werbner, 1998).

Conflict generally enters the picture in these studies as group or ethnic conflict; rarely, if ever, do scholars attempt to connect relational selves[1] in conflict (whether as patterns or scenes) to the context/process of globalization. Perhaps this is a reflection of the instrumental model of identity politics proposed by Baumann (1996) in which culture and community are concepts generated by those with the authority to shape hegemonic discourse and/or to define an administrative order. Against such a backdrop the agency and position of both dominant and oppressed groups is obscured and the discursive operations of power dynamics ignored. Despite (and perhaps because of) the lack of consistency in the use of concepts such as community, culture, and globalization, it is worthwhile to look at these terms as important contexts for the communication of conflict, whether the scenes are viewed as the product of industrialization, globalization, postcolonialism, postmodernity, late modernity, or all or none of these.

This chapter will look at several popular frameworks for conflict analysis and intervention and look at the application of those models to the global context. I compare interpersonal, intercultural and international models for the analysis of conflictive communication and analyze their effectiveness in describing global scenes. To accomplish this task I first discuss conflict and globalization as I am using these terms in the chapter; then, using Levinger and Rubin's (1995) framework, I briefly compare interpersonal and international contexts and their different impacts on the conflict process. Next I discuss several frameworks for analyzing intercultural conflict and the explanatory potential of such models for accounting for globalization processes. Finally, I propose a model of conflict analysis that emerges from the context (here conflict over ethno-national identities on a national listserve group).

CONCEPTUALIZATION OF CONFLICT AND GLOBALIZATION

For the purposes of the arguments posed in this paper, I advance two of the most popular definitions of interpersonal and intercultural conflict. Interpersonal conflict is defined by Wilmot and Hocker (1998) as: "an expressed struggle between at least two interdependent parties who perceive incompatible goals, scarce resources and interference from others in achieving their goals" (p. 34). Building on this definition, Ting-Toomey (1998) defines intercultural conflict as the "perceived incompatibility of values, norms, processes or goals between a minimum of two cultural parties over identity, relational, and/or substantive issues " (p. 401) and notes the importance of studying intercultural conflict in the global scene. Of primary interest to communication scholars is conflict as it is expressed in interaction. This would exclude perceptions of conflict as they exist within an individual's mind and to the extent that these feelings remain unarticulated. For instance, if I say that "I am feeling conflicted," I am expressing a perception of conflict. While it is important to analyze the expressed meanings of conflict, the focus of much communication and other research is on recall through self-report and survey of *expressed struggle*. Although the focus of analysis is on the interaction, the data is still based in perception or general impressions of events. While both the perception (given voice) and the acting out of conflict (in interaction) are equally important, the links between the two are rarely studied as interactive events or stories. It is my contention in this chapter that it is precisely this connection that processes of globalization unmasks, for globalization lays open questions of nations, cultures, groups, communities, and individuals and boundary markers of identity and difference, and thus for expressions of struggle as within or outside of a complete or bounded self(ves).

Globalization transcends the emphasis on communication between states (international communication) to encompass the growing importance of non-state actors (e.g., transnational corporations and non-governmental organizations). There are discrepancies (information rich/poor) between the developed countries (25%) and "third world" countries (75%) in terms of the distributions of communication infrastructure/technology, control over export/distribution channels, and access.

Global communication has been defined as the study of the transfer of (mediated and otherwise) messages between individuals and collectivities. These messages may be enabled or constrained by institutional or structural forces, access to information and communication technologies. In Latin America, Africa, and Asia there is a wide discrepancy between those who control or have access to information and those who do not. In Central America, for example, intergovernmental organizations (e.g., UNICEF, UNESCO), foreign service organizations (e.g., Peace Corps) and Nongovernmental Organizations (NGOs) have attempted to create bridges between "developed" countries (those in control of information distribution) and those groups in the country with less access to information and information technologies. These programs often take the form of educational training (computer, medical, legal, etc.), which can extend, overlap or be substituted by training in health, labor, and social issues (e.g., family planning, domestic violence, conflict resolution, worker's rights). Where many of the programs are "imported" they sometimes fail to consider meanings given priority and importance in the local community, and thus provide little incentive for the often monumental effort it takes to sustain change in the community.

Likewise, studies of global communication rarely analyze the ways in which public spaces are recreated and used by diasporic groups for the purposes of social movement and change. Werbner's (1998) study of the ways Pakistani women's groups used public spaces in London following the Rushdie affair shows that global events do not structure interactions among groups as simply dominant or oppressed, but as multiple spaces for duties, rights, and interests:

> In reality ... people bear multiple conflictive loyalties and quite often—as is the case with British Pakistanis—multiple formal citizenships. . . The public march organised by women in Manchester to protest against

> the atrocities in Bosnia and Kashmir was a demonstration by settler Pakistani women against the policies of their own (British) government in which they highlighted the particular predicament of Muslim women, citizens of India, to whom they felt attached both as women and as Muslims. But the march also asserted the marchers' citizenship in a third country, Pakistan, which had an historical stake in the Kashmir dispute and was, additionally, the women's homeland, a land to which they were attached by familial, sentimental and patriotic ties. The attachment was one of honour as well as of sentiment. (p. 24)

In these contexts, it is helpful to analyze global communication not only as the study of message flow on a structural level, but as the ways in which group, community, and national identities emerge and come into conflict. *Globalization* then is both structurally defined and emergent in interaction, in the creation and sustaining of global entities and identities, and in the displacement of nation, culture, and community. If globalization can be seen as the compression of culturally salient markers of time and space and as the displacement of territory and state as grounds for national identity—as ethnic, immigrant, and diasporic identities are reinventing and retelling national histories—then models of conflict at the interpersonal, intercultural, intergroup, and international levels need to be reconfigured to account for these changes. Conflict, in this context, cannot be analyzed on the basis of individual, ethnic, cultural, or national identities as if such identities are predetermined and thus determine the course of conflict. This is not to imply that identities which were once whole are now fragmented and incomplete, due to the (unequal) travel of information technologies from the West to the rest. Rather, the categorizations of conflict analysis and research as international or intercultural are now displaced by the new histories as geographic and disciplinary boundaries are reconfigured.

Nonetheless, it is important to determine the potentials and limitations of interpersonal, intercultural, international, and interethnic theorizing as we move toward a conceptualization of the communication of conflict in the context of globalization. In the following paragraphs I review some basic models for understanding and analyzing conflict in cultural and national contexts.

FRAMEWORKS FOR ANALYZING INTERPERSONAL, INTERNATIONAL, AND INTERCULTURAL CONFLICT

Levinger and Rubin (1995) propose that, while theorists had for some time regarded conflict processes through the lens of their own disciplines (mathematicians using game theory, for instance), books such as

Deutsch's (1973) *Resolution of Conflict*, and Boulding's (1962) *Conflict and Defense*, among others, influenced scholars to think of all conflict processes as fundamentally alike. Levinger and Rubin are quick to point out that, indeed, while there are some similarities, it is possible to differentiate among the contexts in which conflict takes place. They go on to point out the similarities and differences among interpersonal and international contexts for conflict, making generalizations about each of these contexts in doing so: number of parties, number of issues, exit, power asymmetry, enforceability of agreements, effectiveness of third party intervention, audiences, representative negotiation, information about one's counterpart, trust, institutions, and bureaucracies.

The argument put forward by Levinger and Rubin suggests that while there are some similarities between interpersonal and international conflict, the main differences lie in the complexities that occur when multiple views, bureaucratic, and institutional embeddedness are introduced. While certainly there are very complex differences between the two contexts, their analysis fails to move beyond the structural barriers to look at the actual process of communicating conflict, the interaction itself. Although Levinger and Rubin never indicate an interest in interaction per say, it would seem that the *communication* of conflict in these different contexts would be the focus of such analysis, rather than general structural differences in context. Still, their approach is typical of many conflict scholars who see communication as a tool in the conflict process, rather than as creating the process itself.

Another important distinction made among conflict processes occurs on the intercultural level. There are several frameworks for conceptualizing intercultural communication, as there are for understanding conflict. In the following paragraphs I demonstrate the problems and possibilities with using the covering law approach to understanding intercultural conflict. To first examine the fundamental bases of difference used throughout intercultural studies, I look at Hoppe, Snell, and Cocroft's (1996) use of Fiske's typology. Fiske (1991) discusses communal sharing, authority ranking, equality matching, and market pricing as being fundamental building blocks to coordinating social action. Communal sharing refers to the characterization of relationships as based on group membership. Authority ranking is represented in the hierarchical ordering of peoples in social interaction based on control over resources. Equality matching refers to the perception of equality in relationships and expectations of reciprocity. Market pricing is the evaluation of commodities in terms of the costs and benefits they provide, and judgments about individuals, relationships, and groups are based on such evaluation. Hoppe et al. (1996) map onto

Fiske's typology the four axes of cultural difference proposed by Hofstede (1980): individualistic-collectivistic, power distance, masculine-feminine, uncertainty avoidance.

> Communal sharing processes are primarily collectivistic in that the structure emphasizes the importance of the in-group and of shared group goals. The communal sharing system is feminine in that the relationships in the structure value interdependence and guarding the relationship. The market pricing structure is predominantly individualistic in that the emphasis is placed on individual achieve-ment. In addition, market pricing is masculine in nature because money, achievement and independence are valued. Relationships in market pricing prescribe rules and formal settings, which is standard of high uncertainty avoidance cultures. Alternatively, equality matching is more accepting of difference and ambiguity; therefore, the culture is characterized by a low uncertainty avoidance orientation. (p. 68)

Hoppe et al. continue to add onto the elementary structures proposed by Fiske other dimensions of interpersonal relationships (Burgoon & Hale, 1988) and marriage typologies (Fitzpatrick, 1988). For each dimension of Fiske's model they map out correlations to the dimensions of interpersonal relationships (control, intimacy, composure, formality, task-social orientation, and equality). For example, for the dimension of formality, Hoppe et al., note that authority ranking is characterized by formal roles, and that equality is distributed according to status. Likewise, for marriage typologies, Hoppe et al. note that traditional marriage types are more likely to be strong in authority ranking, low on equality matching, etc. Underlying their argument is the assumption that human social actions are universal and that cultural difference can be mapped onto or explained by differences in the interpretation of each elementary unit.

To continue the lines of the argument advanced by Hoppe et al., I add to their Fiske/Hofstede intercultural model the notion of conflict styles as elaborated by Kilmann and Thomas (1975), among many others. Conflict styles remain the most popular assessment tool in conflict analysis, and thus would have strong implications for intercultural conflict assessment. Kilmann and Thomas discuss five styles: collaboration, accommodation, competition, avoidance, and compromise. Competition is characterized by confrontational and uncooperative behavior, by win-lose tactics, and by establishing power over another. Accommodation, on the other hand, prioritizes the needs of others over asserting one's own needs and is often called the "doormat" style. Avoidance is typified by not acknowledging the presence of a conflict or by running away from conflict. Compromise is the style that allows disputants to "meet halfway," where both parties gain something and both lose something. The preferred conflict style, for those interested in conflict resolution anyway, is collaboration. Collaboration demonstrates a high concern for the self as well as for the other, and demands a high amount of involvement, investment in the relationship, and interaction on the part of both parties.

When these styles are charted onto the Fiske typology, we see that competitive styles might be characteristic of cultures that emphasize authority ranking and market pricing. Accommodation would be a style that is characteristic in cultures that accentuate communal sharing (collectivism) and authority ranking. Avoidance, likewise, would be indicative of cultures that are high in communal sharing and authority ranking. Compromise would be more difficult to predict, although it could show up in cultures that stress equality matching and market pricing to a lesser extent. Collaboration, then, would tend to be a characteristic style for people in cultures that value market pricing first, and then equality matching.

Extending the analogy further one could add goals in conflict to the mix. Wilmot and Hocker (1998) use the C.R.I.P. (content, relational, identity, process goals) model to discuss conflict goals: content goals are goals (during conflict episodes) which relate to what is said; relational goals are goals which derive meaning from the assumed/projected relationship between the parties; identity goals are goals which indicate something about the self that the person (or group) in conflict is interested in preserving; and process goals deal with what communication processes will be used in the conflict.

Assuming the hypothesized correlations discussed above, we could see a basis for making arguments for predictability of intercultural conflict based on the importance and meaning of goals culturally. Content goals would be salient in cultures where the emphasis is on equality matching in relationships, communal sharing is low, market pricing is evident, and authority ranking is not emphasized. Relational goals would be important where communal sharing and authority ranking are emphasized. Identity goals would become salient where equality matching and market pricing are emphasized (although group identity is salient in communal sharing cultures). Finally, process goals would be important in cultures where equality matching is a strong value.

Along the same lines, Ting-Toomey (1998) builds a case for examining intercultural conflict on the basis of differences in conflict models, styles, rhythms, norms, and ethnocentric lenses. Conflict models may be outcome oriented (that is, preferring to assert individual goals and move rapidly toward achieving those goals) or process-oriented, which emphasizes the management of face and context before moving to consider goals. Styles for Ting-Toomey vary from indirect to direct, while norms are described as either equity or communal (similar to those discussed in the model above). Conflict rhythms express cultural orientations to time, which Ting-Toomey suggests influences movement toward clear goals and objectives. Finally, ethnocentric lenses refer to the attributions cultural peoples make about others' behavior based on their own personality or situational biases. While Ting-Toomey discusses cultural differences in conflict in more depth than does the earlier framework, both approach conflict interactions from the basis of individual and group differences and advocate intercultural conflict competence on the basis of learning about those differences.

The frameworks for analyzing intercultural conflict discussed above utilize a hypothetico-deductive approach to understanding cultural difference. Basic units of interaction are identified which are said to underlie all aspects of social interaction. On these "elementary units" other typologies and models for understanding culture and conflict can be built, to indicate categories into which differences can be classified and explained. It is important to note that an analysis of the place of culture in conflict process has generally been approached from the standpoint of differences between cultures (conflict in intercultural communication). Emphasis in studies of intercultural communication and conflict tend to look for some standard by which to measure cultural differences. Cupach and Imahori (1993), reflecting many studies in this area, look at identity as measured and measurable on *levels*—one can proceed or move from an interpersonal to an intercultural level depending on the position of relationship to the other (toward or away from intimacy). Cultural differences then are on a different level from interpersonal differences and thus imply different things about distance, closeness, and intimacy. From this perspective, intercultural conflicts can be differentiated from interpersonal conflicts in the degree of intimacy (knowledge about the *real* identity of the other) implied in the level of relationship[2]. Intercultural conflicts take place among people who do not know each other as well as those who have interpersonal conflicts.

Moving from models which focus on the generalization of micropractices to the macrostructural level, Rabie (1994) develops a Shared Homeland model to account for conflict and ethnicity in globalization. Rabie first recounts changes on the national and global level that have led to new distributions of power and the weakening of national sovereignty. Thus the policies of conflict resolution and (more popular) conflict containment that were politically expedient in the past, bear little relevance in the current global political theater. Rabie notes that, given the collapse of communism and ethnic/national battles in Europe and around the world, there is "global movement toward greater political freedom and democracy, greater respect for human rights, and a quest for economic development and regional cooperation." Rabie suggests several conditions that are necessary to deal with these changes and with the nationality "problem" include:

> ... acceptance of the inviolability of existing state borders, recognition of the need for and benefits of economic integration and regional cooperation, respect for the rights of others to choose their own political and economic systems as well as cultural identity, mutual commitments to democratic pluralism as a way to govern oneself, conduct political dialogue, and deal with the never-ending social and political conflict with others, acceptance of collective security arrangements as a means to enhance mutual independence and as political frameworks to improve prospects for regional stability. (pp. 175–176)

With these conditions in mind, Rabie proposes a Shared Homeland model in which there is political separation along national and ethnic lines while "maintaining or initiating economic and social unity across political lines" (p. 177). This would mean that ethnic groups would be allowed to form autonomous regions and have separate homelands within the larger state and would have equal access to the social and economic services provided by the state. Rabie maintains that such a set up would allow "each minority to have a homeland of its own [and] the combined homelands of all political minorities would form a shared homeland for members of all national groups to enjoy" (p. 177). Historians and others writing on immigration, citizenship, and membership (e.g., Glazer, 1987; Takaki, 1987; Walzer, 1998) have long argued over the tensions that arrangements such as those proposed in Rabie's model are likely to produce. In some limited cases, ethnic or tribal autonomous regions in a "national" space have existed peacefully and to the benefit of the minority group (the Kuna in Panama are one such example). Yet, until the concept of "nation" is no longer relevant in global politics the Shared Homeland model will be politically untenable.

In the confusion among temporalities, spaces, identities, and nation-states, the connection to conflict, social movement, and globalization seems tenuous at best. The construction of the individual as citizen is arguably one of the primary organizing factors of social life; yet, the panopticon of citizenship has become increasingly elusive. For instance,

Shaman (1994) observes that institutional organizations, such as the Department of Patriotism in Belize, further oppress the country's citizens by presuming one definition of nation and nationalism. The connection between the individual and the social has not disappeared; in fact, Gergen (1990) notes that it has grown stronger. Others agree, noting the success of on-line churches and support groups.

However, Melucci (1993) concludes that, with the glut of communication technologies and quantities of information, all aimed toward the individual who must receive, analyze, and nearly always respond to this information flow, the traditional reference points that mediate individual identity and society, (church, family, etc.) have weakened. We no longer can respond with certainty to the increasingly prevalent question: "Who/what are you?" That we are of this or that race, ethnicity, or nation.

While certainly processes of globalization and communication point to differences and similarities between/among cultures, my emphasis is on communication as central to developing ideas about identities, conflict, and culture. If communication is viewed as the process through which people learn to act in their social worlds, then conflict and globalization emerge as relevant concepts through these actions. To be able to analyze and describe these situations then becomes less a product of analyzing variables so that we may be better able to predict variance across cultures and generalize about differences, and more an interest in developing situated knowledges of the dynamics and complexities of conflict. These are not reductive explanations but detailed accounts that position the tensions of local-global, agency-determination, individual-structural, etc. This model is neither structurally determined nor rooted in free-wheeling subjectivity, it is not relativistic or dogmatic but acknowledges these tensions and connects them to material scenes.

In the next section, I borrow from Melucci's (1993) model of modern social movements to present a case for describing the processes conflict, communication, and globalization in similar ways. My interest here is less about developing a model that "captures" the essence of this conflict than identifying what such a framework might help us to do. My explanation of the framework is followed by a brief example and analysis.

THE MODERN DILEMMAS OF CONFLICT AND SOCIAL MOVEMENT

Melucci (1993) notes that four dilemmas characterize contemporary social movements. I will discuss each dilemma briefly and then look at their relationship to conflict and globalization and to related arguments regarding democracy and cyberspace. The first dilemma can be characterized as that of autonomy and control—"between the potential of individual capacities and choices versus the tendency to create capillary systems of behavioral manipulation, which today impact both the brain and the genetic structure itself" (p. 290). This tension can be found in the arguments for and against historicity and subjectivity and the creation of imagined communities, and in the related controversies over the recognition of agencies and abilities to negotiate identities verses the always already determined "nature" of those actions. Here, the potential of individual capacities and creative choices is balanced by the (increasing) power to create and control the "virtual" subject.

This leads to Melucci's second and third dilemmas, which I see as somewhat similar: that of responsibility and omnipotence and irreversible information and reversible choices. With these dilemmas, the tension lies "between the tendency to expand the capacity of human systems to intervene in their own development versus the need to take account of and respond (response-ability) to the limits of nature" (p. 290). Melucci's ideas are based on the power of scientific knowledge and the power of choices that must be acknowledged. Although he is discussing science as power, the same tensions can be related to the choices made in representing cultures and nations.

Melucci's final dilemma, that between inclusion and exclusion, is the tension of defining and defying difference in contemporary society. Melucci notes that, due to advances in technology that collapse time and travel, there is no longer an "outside world": "territories, cultures, languages exist only as the internal elements of a planetary system" (p. 290). Thus differences are internalized as inclusive, i.e., processes which cover over and level identities as equally fragmented according to the cultural codes of the dominant elite or exclusive, (i.e., processes which resist the homogenization or silencing of cultural forms).

The usefulness of this framework for explicating the dynamics of contemporary social movements can be seen in how well it addresses both the macrosocial structures that frame cultural contexts as well as the micropractices that negotiate and recreate meanings within the larger framework. The application to conflict processes in the context of globalization seems equally clear. The invocation of autonomy suggests that one has the ability and agency to exhibit free choice and to be free from situations that oppress one's identity (however that may be defined). Yet control over information and information technologies means control over knowledge, histories, and mediums of disseminating information. Conflict emerges in the tensions between transnational conglomerates and first world nations and their governments and between those entities and "third world" countries over control of resources increasingly defined by their informational capacity. Conflict over the meaning and nature of sovereignty in the global theater is likewise reflective of this dialectic.

Still, these dilemmas can exist simultaneously and in numerous spaces. In a different context, autonomy could represent personal choice and personal freedom to express one's identity, to choose to be a citizen, to choose to travel, or to migrate. Tension here emerges over who chooses travel, sojourn, or emigration and who is exiled—between those defined as immigrants and ex-patriates and all of these categories in relation to the classification of citizen. These concepts are covered in more detail in Melucci's later writing (1997) on the production and reappropriation of meaning and on processes of *individuation* from multiple selves as necessary to forming autonomous and responsible subjectivities.

These tensions spill over into the second and third dilemmas of omnipotence and responsibility and irreversible information and reversible choices. This dilemma is perhaps even more frequently represented than the first in the omnipotence of those who control the means to information to control the "rest" of the world, and the responsibility implied in that control and the acknowledgment that such information irrevocably alters the global system. The tensions between the creation of systems that regenerate themselves and choices to not intervene or violate "natural" systems seems to be with those nations who have access and control of information. Still, the relationships among (first and third world) nations and transnational corporations are more complex and interdependent than national borders and sovereignty would indicate, and global power is becoming more discrete and less reliant on the dominance of any one national state.

The fourth dilemma, that of inclusion and exclusion, overlaps and implicates the others. Identities in conflict presume and preserve certain meanings of self and other even as they challenge essentializing tendencies in global confrontation and change. The meanings for identities included or excluded in any conflict are part of a material scene that may not be physically or geographically bound. In global contexts such as internet usage, worldwide media consumption of Western programs, multinational commerce and bloc trading, what is included or excluded is constantly displaced; locations shift as identities are differentially bound by (physical, cyber) spaces and geographical locales. This dilemma is illustrated in more detail in the example discussed below.

NATIONAL IDENTITIES, CYBERCOMMUNITIES, AND GLOBALIZATION

One of the most accessible sites for scholars to study the process of globalization, conflict, and communication is not in physical boundaries fought for or erased in spaces of nations or land claimed through ethnic histories, but the link between those boundaries and the imagined communities constructed along these lines in cyberspace. I begin with a message from a Panamanian listserve (Panama-L) that followed a thread of debate over who could/should participate both in the action to stop the government's construction of a highway through the heart of the metropolitan park and rain forest in Panama City.[3] The critique of that action and *who* should have a voice reached an interesting point with the distribution of a letter that addressed neither of these issues, but, rather, the construction of Panamanian identity in both the U.S. and in Panama:

> My name is Gregorio. I was born at Gorgas hospital in Balboa and raised between Pedro Miguel (in the canal zone) and Las Savanas (directly next to the Panamerican Institute)... I have beenliving in Seattle for 12 years this August 6. I keep in touch with most of the Panamanian population here in Seattle and we try to do things together whenever possible ... Without meeting any of you, I have already begun some controversy within the group which was not my intention [referring to his earlier comments about living in the US v. living in Panama]. . To those of you that might be slightly irritated with me for entering the question or others for their comments, I give you my sincerest apologies and ask to restore the peace like it once was. As people of color we face too much conflict from some African Americans that have a problem with someone who is a darker shade of brown and has such ties with the Hispanic community; and at the same time with some in the Hispanic American community for being born in Latin America and being a darker shade of brown. Please do not misinterpret what I am saying. Let's just not fight with each other. . .
>
> Talk, Dark, and Panamanian
> Gregorio (Springer, June, 22, 1995)

Gregorio's "letter" followed an initial "call to action" to stop the construction of the highway. However, it is his call for community in particular that I wish to focus on here. Although this letter did not register direct criticism of the highway construction, it addressed the conflict that had been going on among the members of the site, while simultaneously pointing to the conflictual representations of Panamanian identity in the United States.

While the earlier messages[4] had each constructed national space geographically (constructing legitimate participation in the debate on the basis of the experience of either *being* Panamanian or *living* in Panama), Gregorio reconfigured this space and his place in the controversy with an interesting move. He first replicated the earlier criteria for validating authenticity: he noted the exact locations of his birth and early years within the national space of Panama. Then he discussed his frequent contact with and ties to a "displaced" Panamanian community in the U.S., a community that is continually marginalized even by those at the margins of mainstream U.S. society (namely, Latin Americans and African Americans). Thus the primary story for Gregorio was the "living out" of Panamanian identity in spaces where he did not feel marginalized but rather a part of a community. This virtual space for him was not a space for debating *being* Panamanian, but rather a space where his Panamanian identity can be confirmed and supported. He thus constructs this virtual space as a place for a different kind of national identity and constructs a community of individuals united in their differences from others.

The thread of discourse that preceded Gregorio's message presented the highway action as a measure designed to suppress information (thus preventing free choice) about the destruction of an important natural habitat (issue of control). The dilemmas of responsibility (as a Panamanian citizen or expatriot) and reversing the choices made by the President thus placed a national issue in a global forum (a listserve with subscribers from all over the world).

The thread that followed the call for action to stop the highway construction did not address the same tensions (choice, responsibility, and control) around corporate greed and threats to the environment—or the action itself for that matter—but instead attempted to place boundaries on just who could or should be able to criticize the movement to stop the construction of the highway. The responsibility at this point seemed to lie more with representing oneself as authentic and one's interests as aligned with those supporting the action. Disagreement was allowable, but only if one first met the criteria for engaging in debate about the future of the country. Gregorio obviously disobeyed these rules by critiquing first and providing justification (via identity) for his critique after he had been chastised by the group.

Thus, the implication of the inclusion/exclusion dilemma, which here functions on a multitude of levels. First, as mentioned above, the criteria for providing support or critique served a function of boundary maintenance. Inclusion in the listserve group was the primary dilemma for Gregorio, given the exclusion he faced elsewhere. On a different level, for Gregorio, inclusion and exclusion among the African American and Hispanic groups in Seattle seemed to be based on skin color—and Panamanians, according to this criteria were too dark to "fit in" to either category. Lastly, inclusion or exclusion functioned on the level of strategic essentialism. It was important for Gregorio to identify certain qualities as essentially Panamanian (e.g., language, skin color) in order to call for "peace" among the members of the group.

Within the theorizing on democracy and cyberspace these dilemmas also can be easily spotted. Discourse that portrays technology as the means to democracy carries with it the equally weighty implication that technology/democracy provides "citizens" with a forum for justice (social, political, and in some cases, economic). Thus, all citizens have equal access to change policies and with this access, have the freedom to "represent" any number of differing identities. Beyond the fact that only 3% of the world's population have access to the internet (see Sassen, 1996, for a breakdown of these numbers), the inclusion argument thus positions opposition to technology-as-equality as anti-democratic. Other dilemmas, such as autonomy and control and responsibility and omnipotence, are the tensions fundamental to the creation and maintenance of democracy in late modern society. Traditional Marxist arguments about the (lack of) access to technology, also align with these binaries, though starting with the exclusion (as opposed to the inclusion) argument and then working to address the other tensions. The focus of the Marxist arguments thus center around: (1) the failure of democracy to include the people; and (2) the relationship between control over electronic space and the flow of transnational capital. Thus, the responsibility of the (disenfranchised) people is to unite against the global (omnipotent) corporate interests and fight against the commodification and consumption of individual and national identities.

IMPLICATIONS FOR CONFLICT ANALYSIS: DIALECTICS AND DIALOGUE

Working from these dilemmas and the analysis to the idea of conflict in the context of globalization, it seems that these tensions are present in the discourse. But do they adequately account for the complex negotiations necessary to track the movement mentioned earlier, from historicity to subjectivity, from the local to global? Can they/should they provide a perspective on the disjunctive forms of representation that signify a people? Chambers (1998) notes that:

> We all seek to recapture plenitude, the wholeness of things, turning away from the unfinished, what, to cite Adrienne Rich, stands unfinished in our accounts. We all desire this coherent picture and cognitive mapping, but perhaps there also exists the way and necessity for registering the limits of such imaginings; of permitting what we turn away from—the always unfinished world we inhabit—to insist on its right to be. This would register what cannot be immediately quantified and defined, what resists the technical representation of reality, and yet insists and continues to interrogate us. (p. 44)

I believe that these tensions are a useful beginning for locating a dynamic that needs a more complex reading, incorporating all the factors cited above, but employing none as a justification or validation for analysis. While it can be useful to look at complications between interpersonal conflict and international conflict in terms of representation of issues and formalization/flexibility of rules, these models and comparisons do not adequately describe/reflect the process and material global scene. Likewise, models for understanding cultural differences and conflict styles are heuristic and predict patterns and tendencies among groups in conflict, but do not generate much useful knowledge about specific scenes nor locate those scenes in the context of global communication. Moreover, neither of these approaches addresses the ways communication positions interactants discursively within these scenes and the power effects of such positioning. Rabie's (1994) Shared Homeland model is derived from natural ethnic conflicts around the world and recognizes underlying tendencies among nations and ethnic groups as a result of the process of globalization. Yet, Rabie's model ignores the microdynamics of how/why conflict actually occurs to focus on what is necessary to resolve conflicts on the macrolevel. His model, while hopeful, ignores the historical tensions between membership and citizenship. What is lacking in both approaches to conflict and globalization is a perspective that connects theorizing about globalization as a process with application to the local-global nexus. If what some critics of the "epidemic" of global theorizing (e.g., Beng-Huat, 1998; Hirst & Thompson, 1996) propose is true:

> (1) that much of ethnic conflict has its origin in local conditions and can be explained by local histories, (2) that the bulk of foreign direct investments come from Japan, USA (known as the interlinked economies [ILE]) and Europe and are directed toward their respective spheres of influence, (3) that while multinational companies are geographically spreading themselves, the spread is essentially restricted to the ILE and some countries in East and Southeast Asia, (4) that more than 70 percent of the activity of international business is home-based

> or nationally embedded and (5) that the globalization of capitalism will afford poor countries the opportunity to acquire technologically advanced industrial economies within the next 25 years is fanciful (Beng-Huat, 1998, p. 121)

then analyses in this area need to pay attention to the interdependence of the local and the global within specific sites as nested in macrolevel dynamics. As Beng-Huat observes, few globalization theorists would go to the extremes of the arguments presumed above; however, most would agree that the unevenness of the spread of the world's resources is an issue to be analyzed and its implications exposed, "rather than reaffirming the status quo as a manageable entity," as some critics would do (p. 122).

While the tensions proposed in Melucci's model do not finally provide an explanation of these dynamics they do account for the communication of conflict as process and they do allow for connections between the local and the global. Still, to the extent that Melucci's model could be characterized as dialectical, it presents relationships as dilemmas—contradictory and oppositional. Such theorizing flattens out the temporality and spatiality of interactions, conversations, and identities in these contexts. As Hawes (1998) observes, "The installation of an individual, as a conscious subject, into the realm of language is, in large measure, the marking of identity and difference, of presence and absence, of sound and silence of self and other. In the realm of language, an individual is alternatingly and often simultaneously, subject and object" (p. 274). Hawes advocates for dialogics over dialectics, noting that dialogically organized conversations, "are both contradictory and oppositional as well as heteroglossic and unfinalizable" (p. 274). In this vein it is perhaps useful to note that Melucci's dilemmas have been called both dialectic *and* dialogic by those writing on conflict and social movement (e.g., Werbner, 1998).

Still, one might ask what is assumed when dialogue and dialogical methods are advocated in uneven and heteroglossic global spaces. Scholars discussing the use and abuse of "dialogue" as a means to conflict resolution among racial, ethnic, and cultural differences (e.g., Du Bois, 1903; Ellsworth, 1989; Freire, 1970; Hooks, 1984) note that dialogue has only happened under conditions of white supremacy and/or conformity to the conditions and structure (for dialogue) of the dominant group. Moreover, conflict resolution "talk" toward racial and cultural understanding in the United States is often structured within relations of difference (as noted in models above) and based on hegemonic constructions of argument/debate and dialogue. Where argument and dialogue are positioned as oppositional terms, analysis of the Western tradition of debate and polarization of issues reveals the fundamental belief that the presentation and defense of position leads to dialogue and conversation. Such beliefs are rooted in beliefs about equality in presentation and representation of self (as non-gendered, non-raced, non-classed, etc.). Here, if one presents oneself effectively, advocating for representation and position, one is engaging in dialogue.

In the context of global communication, those who analyze conflict must increasingly adopt a perspective which refuses the characterization of dialogue or debate as a means toward the *effective management* of differences, but rather studies the ways in which people discursively construct their participation and are constructed in conflict episodes. The politics of domination and oppression and of unevenness in travel across global spaces does not disappear with such a view but is placed within the affordances and constraints of human communicative activity.

Bhabha (1990) describes the space of the people as both the historical "objects" of a nationalist pedagogy, giving the "discourse an authority that is based on the pre-given or constituted historical origin or event," as well as the "subjects of a process of signification that ... demonstrates the prodigious living principle of the people as that continuous process by which the national life is redeemed and signified as a repeating and reproductive process" (p. 297). In the contexts presented in this paper, globalization can be seen as a set of contexts in which local senses of place are confused with global bodies and spaces, where identities are fragmented and contested but not completely determined either by technology or citizenship. The question for communication scholars remains as to what types of approaches to describing, understanding, and analyzing conflict in this setting are useful, to whom and for what purposes. This chapter has attempted to identify some of the boundaries of previous models and to propose a framework for extending, rather than reducing, the complexities of conflict and globalization processes.

NOTES

1. By relational selves I am referring to the self that is presented in-relation-to another—the communication of relationship, identities, communities, and culture in situ. I differentiate this approach from most conflict and communication research which focuses on the individual and group perceptions of difference.

2. According to such a view, intercultural conflicts would take place among people who have not attained the interpersonal level of intimacy.

3. This call to action on the listserve is elaborated and analyzed elsewhere (Cooks, in press).

4. A more in-depth analysis of this call to action and the construction of national identities on this listserve can be found in Cooks (in press).

REFERENCES

Baumann, G. (1996). *Contesting culture: Discourses of identity in Multi-ethnic London*. Cambridge: Cambridge University Press.

Beng-Huat, C. (1998). Globalisation: Finding the appropriate words and levels. *Communal Plural*, 6: 117–124.

Bhabha, H. (1990). DissemiNation: time, narrative, and the margins of the modern nation. In H. K. Bhabha (Ed.), *Nation and Narration* (pp. 291-321). London: Routledge.

Boulding, K. (1962). *Conflict and defense: A general theory*. New York: Harper Torchbooks.

Burgoon, J. K. & Hale, J. L. (1988). Validation and measuement of the fundamental themes of relational communication. *Communication Monographs, 54*, 19–41.

Chambers, I. (1998). A stranger in the house. *Communal Plural,* 6, 33–50.

Cooks, L. (in press). Negotiating national identity and social movement in cyberspace: Natives and invaders on the Panama-L listserve. In J. Ebo (Ed.), *Cyberimperialism* (pp. 1–14). Westport, CT: Greenwood Publishing.

Cupach, W. R. & Imahori, T. (1993). Identity management theory: Communication competence in Intercultural episodes and relationships. In R. L. Wiseman & J. Koester (Eds.), *Intercultural communication competence* (pp. 112–131). Newbury Park, CA: Sage.

Deutsch, M. (1973). Conflicts: Productive and destructive. In F. E. Jandt (Ed.), *Conflict resolution through communication*. New York: Harper and Row

DuBois, W.E.B. (1903). *The souls of black folk*. Greenwich, CT: A Fawcett book.

Ellsworth, E. (1989). "Why doesn't this feel empowering? Working through the repressive myths of critical pedagogy." *Harvard Educational Review, 59*: 297–324.

Fiske, A. P. (1991). *Structures of social life: The four elementary forms of human relations*. New York: Free Press.

Fitzpatrick, M. A. (1988). *Between husbands and wives: Communication in marriage*. Newbury Park, CA: Sage.

Freire, P. (1970). *Pedagogy of the oppressed.* New York: Herder and Herder.

Gergen, K. (1990). *The saturated self: Dilemmas of identity in contemporary life*. HarperCollins: Basic Books.

Glazer, N. (1987). The emergence of an American ethnic pattern. In R. Takaki (Ed.), *From different shores: Perspectives on race and ethnicity in America* (pp. 11–37). New York: Oxford University Press.

Hawes, L C. (1998). Becoming Other-wise: Conversational performance and the politics of experience. *Text and Performance Quarterly, 18*, 273–300.

Hirst, P. & Thomson, G. (1996). *Globalization in question: The international economy and the possibilities of governance*. Cambridge: Polity Press.

Hofstede, G. (1980). *Culture's consequences: International differences in work-related values*. Beverly Hills, CA: Sage.

Hooks, B. (1984). *Feminist theory: From Margin to Center.* Boston: South End Press.

Hoppe, A. K., Snell, L. , & Cocroft, B. (1996). Elementary structures of social interaction. In W.B. Gudykunst, S. Ting-Toomey & T. Nishida (Eds.), *Communication in Personal Relationships Across Cultures* (pp. 57–78). Thousand Oaks, CA: Sage.

Kilmann, R. & Thomas, K.W. (1975). Interpersonal conflict-handling behavior as reflections of Jungian personality dimensions. *Pyschological Reports*, 37:971–980.

Lebesco, K. (1998). *Revolting bodies? The on-line negotiation of fat subjectivity*. Unpublished dissertation. University of Massachusetts, Amherst.

Levinger, G. & Rubin, J. (1995). Bridges and barriers to a more general theory of conflict, *Negotiation Journal, 10,* 201-216.

Loomba, A. (1991). Overworlding the third world. *Oxford Literary Review*, 13, 164–191.

Melucci, A. (1993). The global planet and the internal planet: New frontiers for collective action and individual transformation. In M.

Darnovsky, B. Epstein, and R. Flacks (Eds.), *Cultural politics and social movements*. Philadelphia: Temple University Press.

Melucci, A. (1997). Identity and difference in a global world. In P. Werbner & T. Modood (Eds.), D*ebating cultural hybridity* (pp. 58–69). London: Zed Books

Mitra, A. (1996). Nations and the internet: The case of a national newsgroup, 'soc.cult.indian.' *Convergence*, *2*(1), 40–75.

Rabie, M. (1994). *Conflict resolution and ethnicity*. Westport, CT: Praeger.

Sassen, S. (September, 1996). *The topoi of e-space: Global cities and global value-chains*. Paper delivered at the DEAF symposium, Digital Territories. Rotterdam.

Shaman, A. (1994). *Thirteen chapters of a history of Belize.* Belize City, Belize: The Angelus.

Springer, G. (June 22, 1995). *Message sent to members of the Panama-L listserve*.

Takaki, R. (1987). *From different shores: Perspectives on race and ethnicity in America*. New York: Oxford University Press.

Ting-Toomey, S. (1998). Intercultural conflict competence. In J.N. Martin, T.K. Nakayama, & L. Flores (Eds.), *Readings in Cultural Contexts* (pp. 401–413). Mountain View, CA: Mayfield.

Walzer, M. (1998). Membership. In D. Jacobson (Ed.), The immigration reader (pp.343–364). Malden, MA: Blackwell Publishers.

Werbner, P. (1998). Diaporic political imaginaries: A sphere of freedom or a sphere of illusions? *Communal Plural, 6*(1), 11–32.

Wilmot, W., & Hocker, J. (1998). *Interpersonal Conflict* (5th edition). Boston: McGraw Hill.

CHAPTER FIFTEEN
Listening Across Diversity in Global Society

William J. Starosta
Howard University
Guo-Ming Chen
University of Rhode Island

> All languages of heteroglossia ... are specific points of view on the world, forms for conceptualizing the world in words, specific world views, each characterized by its own objects, meanings and values. As such they may all be juxtaposed to one another, mutually supplement one another, contradict one another and be interrelated dialogically. As such they encounter one another and co-exist in the consciousness of real people . . . these languages live a real life, they struggle and evolve in an environment of social heteroglossia.
>
> — Bakhtin

INTRODUCTION

When persons or nations interact, their exchange is sometimes studied as a consecutive transmission of messages. Such a view supposes that messages have a certain integrity, that such exchanges are structured into sender and receiver, that the message sent must equal the message received, and that receivers play the relatively passive role of accurate processor. An alternative formulation, one that reflects a paradigm shift toward process and the study of a "realm between," is that interactants share interpretive space as listeners, and that their messages mutually shape and reshape the other's transmissions until the interactants reach a point of common satisfaction with the result. Communication then becomes a search for an equilibrium. This chapter adopts this latter view, wherein interactants negotiate fluid meanings using multiple messages, and create an interpretation that belongs to neither party, yet belongs to both of them. This view favors the process of "active listening" over that of mechanical message reproduction or of working toward a predetermined end product. The present analysis offers elements of a listening framework and applies them to a global setting. A listening-centered model of global communication is established.

A MODEL OF GLOBAL LISTENING

Our equilibrium view of global listening is illustrated by means of a model:

Figure 1. An Equilibrium Model of Global Listening

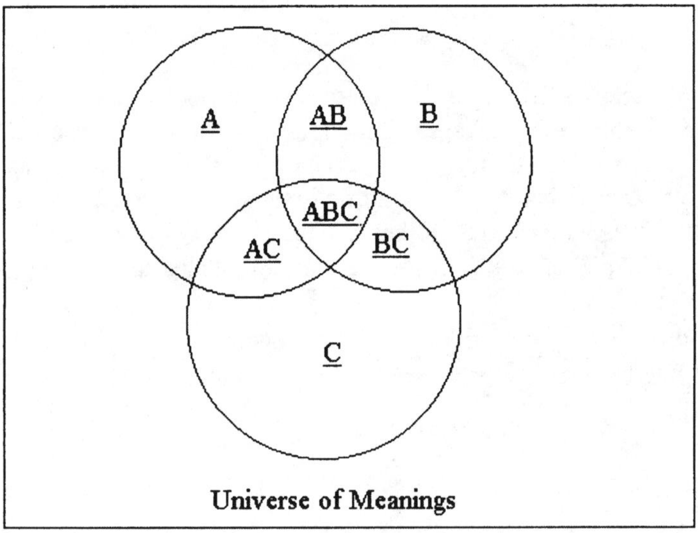

Key: A, B, C are primary message and interpretive communities' meanings. AB, AC, BC are interactions that cross primary community identities. ABC are taken-for-granteds and shared interpretations and experience.

In the model, three interlocking sets are placed within an open-ended universe of meanings. The A, B, and C subsets represent interactions among those of the three respective, primary cultures. Each of these sets provides a derived repertoire of interpretations of the world.

Through the process of interacting the interlocutors negotiate mutual grounds for meanings (AB, AC, and BC, respectively). They also share some taken-for-granteds which make their exchange possible (ABC).

Ethnomethodologists call these assumed commonalties "glossings," while rhetorical analysts speak of myths, topoi, icons, or fantasy themes. To the extent that A, B, and C represent distinctive cultural perspectives and linguisticalities, AB, AC, and BC comprise instances of intercultural listening (Chen & Starosta, 1998).

The listening model portrays AB, AC, and BC as ongoing sites of active intercultural interpretation. The potential for listening to distort messages from urban agencies to rural areas, from masses to élites, from those of one ethnicity to another, and from one nation to another, underpins the present analysis. At each intersection of message sets, listeners may introduce accidental or deliberate message distortion, and may choose to interrogate or not to interrogate suspicious meanings.

The proposed model of listening is heuristic insofar as it looks beyond message transfer-oriented models such as information theory to study the outcomes of a process of mutual dialogue. That such dialogue might be acted out in the presence of power differences and other social stratifications is evident from the author's analysis of co-cultural listening and of third culture building (Chen and Starosta, 1998). Our model proposes that Global Society is found in a negotiated semantic space between, rather than within, existing cultures.

ELEMENTS OF GLOBAL LISTENING

The following section discusses the four elements of global listening based on the model. The elements include: (1) the economics of listening, (2) listening within Culture One versus Culture Two, (3) barriers to global listening, and (4) double-emic listening.

The Economics of Global Listening

The communication discipline fragments knowledge, assigning meaning questions to rhetoric, technology questions to mass communications, accuracy and ethical questions to philosophy, message processing questions to psychology, the expressive function to linguistics, and message selection and distribution questions to journalism. This chapter proposes a holistic communication perspective that centers on global listening. Global listening requires the presentation of self to the messages of the other in an ongoing effort to derive shared meanings.

In at least five ways one account is rendered easier (or more economical) to process than some rival cultural account. A listener generally exercises an easier option over a difficult one, other motivation being equal:

1. Stories told in a familiar language and idiom are cheaper to acquire and process than those that require the learning of a new language, or the reliance on a translator.

2. Stories that harmonize the listener with social norms and conventions are cheaper to acquire and process than others that invite social censure by the receiver.

3. Stories that valorize the listener are cheaper to acquire and process than those that require self-criticism or self-denigration on the part of the receiver.

4. Stories that sound analogous to events already understood are cheaper to process than are stories with no apparent connection to one's own life events.

5. Stories that are consistent with what is already known are cheaper to acquire and process than those that require the creation of a new set of premises or assumptions or world views.

The principle of economy dictates that what the listener hears of the world is constrained by what her narrative framework selects or privileges. Narrative frameworks are assumptive structures that explain what goes with what, what causes what, what should lead to what, what is worthy of what, what counts as a benefit or loss, what end states are desirable. Global listening is enhanced by offering narratives with common language, common values, common ideology, common assumptions, and familiar themes.

Culture One Versus Culture Two

Culture One: The Birth Culture.

> Through a genuinely open engage-ment the strangeness of the other promises to shake us out of our self-certainty, indicating limits where we thought previously there were none, and potentiality where there had previously been limits that defied transgression. (Huspek, 1997, p. 1)

Persons are born into a primary culture from which they draw a world outlook, a preference for styles of living, plus, regrettably, an inclination to enact historical animosities toward those of selected other nations, communities, or groups. This culture of early socialization provides ready answers to common problems, support for ordinary decisions, and an anchor from which not everyone will drift later in life. The cultural pool of symbols and language, broadly-shared meanings and overlapping values and beliefs presents a backdrop against which experience is understood and preferred. Shared psychological or geographic primary identity is considered as Culture One.[1]

Culture One provides a code book for a particular community. Their interpretive community may be defined geographically, or it may represent a geographically-dispersed, yet psychologically-unified collectivity. That is, persons who are separated geographically may still share defining elements of a single birth culture including language, religion, and notions of common origin and ideology. The redundancy of messages and symbolization in Culture One provides a barrier to interaction with those who would introduce messages about other cultures. For those who do not travel or interact beyond Culture One, it may seem a fact of nature that Culture One knowledge suffices for all interactions of any importance.

The strength of Culture One, its ability to closely define a community's identity, becomes also its weakness: Culture One promotes a false sense of self-sufficiency and certitude. Uncritical members of the interpretive community believe that they need not seek out others; others will come to them. They seem to need no new inputs; their closed system will require no effort on their part to adapt to difference. They suppose their values to be universally held by all reasoning persons; only those who are exotic or in error could possibly subscribe to competing values and beliefs. Their isolation, xenophobia, and ethnocentrism appear to them to be natural and desirable conditions. They feel no pressing need or urgency to cultivate global listening skills.

The anchoring of a person in Culture One produces a basis for the negative judgment of others, while it limits community members from learning survival skills for an open, international system. A belief in the moral and physical sufficiency of the birth culture places the community at risk within the larger cultural milieu. When a dominant culture believes in the sufficiency of Culture One, it often acts with intolerance and with discrimination toward other communities and co-cultures. Similarly, it may render other groups invisible, or consider them backward and inferior. In this way, colonial nations usually failed to recognize the cultural achievements of the colonized: diaries of colonial administrators contained abundant references to the supposed primitiveness of those who were colonized, while a generation of researchers in development communication proposed that newly-independent nations should elect a linear path to become like their colonizer so to reach a point of higher civilization.

Culture One rises from truisms, i.e., from beliefs that are redundant, omnipresent, and rarely challenged within a defined, semi-closed system. To that degree, Culture One does not teach how to deal with difference, and the person who challenges the common outlook is outcast or branded as a threat. Community members see little need to rehearse answers for those who would question their outlook; they come to believe these answers to be written in nature and self-evident. Yet, research suggests that truisms can be changed under some conditions: listening across diversity provides a way out of Culture One.

We have proposed that listening across cultures be viewed as a matter of sharing stories or narratives about views of, and explanations for, the everyday world (Chen & Starosta, 1998). The stories to be shared are the facts of Culture One, namely, the accretion of motives, interpretations, values, explanations, and folkways of the birth culture.

So, too, media personnel operate from a Culture One perspective.[2] Their news accounts selected for retelling those dramatic features of stories that fit most closely what their readership is inclined to accept (see Bormann, 1972, 1980, 1982 on "rhetorical visions" or "fantasy themes"; Starosta & Hannon, 1997), while keeping the story interesting. For a Culture One audience, only certain narrative accounts potentially prove plausible and interesting; and, when a Culture Two is introduced that includes persons with other experiences, cultures, and predilections, the same narrative no longer proves compelling. Explains Farrell (1993), "There is always a strain between the world in which discourses appear, and the interiorized subjectivity that gives meaning to the discourse itself" (p. 149).

News reporters have been considered "naïve ethnographers." They listen to accounts and study events, and then report for some a (presumably) fair-minded reduction of events. Whatever these events, each interpreter experiences events through personal filters. Different reporters who "read" a common historical "text" arrive at personal readings of that text. The character who is "good" in one reading becomes the "villain" of another reading of ostensibly the same historical text. The extraction of rival readings from one set of historical "facts" illustrates the multilexicality of "polysemous" texts (Barthes, 1971; Fletcher, 1992, p. 135; Starosta & Hannon, 1997).

Culture Two: Communication across Diversity.

> A dominant group must decide what reading to make of narrative accounts of non-dominant communities, choosing amongst responses that range from (1) actively welcoming the competing account as an example of diversity; (2) merely tolerating the different account; (3) rendering the discrepant narrative invisible; (4) teaching "the truth" about the "mistaken" accounts recounted in the discrepant narrative; (5) converting those to "truth" who discover ironic readings of mainstream

> texts; (6) castigating those who believe in a way that the dominant group does not sanction; or (7) taking decisive, violent action against those whose readings of texts differ from their own. In other words, the treatment of local persons who tell variant narrative accounts of events from those sanctioned by the dominant community ranges from the enjoyment of a good alternative story, on the one hand, to the elimination of the "false" storyteller, on the other. (Starosta & Hannon, 1997, p. 144)

When conditions arise that require a member of one community to listen to those of other groups, a common expectation is that others subscribe to the same values and outlooks as do those of Culture One. This tendency to recognize no difference from themselves on the part of cultural elites has been criticized by post-modernists as a means of silencing opposition, of maintaining hegemony and of seeking to impose one model of development on persons of differing origins and predilections. At a given moment, members of national or world co-cultures may be acutely aware of difference, power relationships, and the need to preserve their culture against the wishes of the other. Mostly, though, it is those of minority cultures (indeed, Orbe treats only such minority cultures as "co"-cultures[3]) who adjust their communication strategies to accommodate such difference.

When persons of differing birth cultures or co-cultures do engage in communication, they dance between an assumption of identity and the fear of difference. At the one moment they express relief when differences do not seem as great as feared; at the next, they wonder if there exists any way to get an understanding across to the other. Mutual unfamiliarity with the other's Culture One compels Culture Two interactants to seek a common ground, if they have the fortitude to remain in contact at all.[4]

In those cases where Culture Two interactants have had historical dealings, sometimes at an unequal level of status and power (e.g., Israelis and Arabs, British and Nigerians, African and European Americans, Protestant and Catholic Irish), historical relations among those of the two communities become an implicit context for the given message. One point is spoken, while some other is easily read between the lines. Does the White interactant suppose a Black counterpart silently holds him or her accountable for slavery? Does the Christian silently consider a Jewish interlocutor as racially responsible for the death of Christ? Do Indian Hindus silently think of Indian Muslims as insisting on rights and privileges that are not available to the majority of the nation? What other premises are tacitly "heard" by one or another partner to an interaction among members of historically-related communities? Can these silent premises effectively deafen a listener to the meanings and intentions of the other? Attributions that occur during interaction across diversity may introduce the "noise" of unspoken premises as an interpreter of observed behavior.

Those of different birth cultures learn to employ a style of communication that is comfortable and natural to them. This style may differ very considerably from the style of those of other cultures. One group may practice an oral style, while another prefers writing. The first, including many Native Americans, may prefer personal testimony to demonstrate truth, while the latter, including many European Americans, may seek documentary expressions. Southern orators of American history or contemporary Arab spokespersons may be inclined toward ornate expression, whereas others, such as Koreans or Japanese, may generally prefer fewer words. One group may favor self-disclosure and assertiveness, while another values privacy and reserve.

Listening across stylistic difference may lead cultures to misjudge one another as deceitful or threatening, as sly or hostile, as impolite or inappropriate, as self-serving or verbose, as weak or irrational. The form of a message often interferes with the ability to process the content: How much of the message is to be taken as exaggeration? How sincere is the position of the interactant? Is the communicator hiding some intent, is something being said while not being said, and what is the reason for such attributed non-disclosure? Should bold statements be read as content or an indicator of sincerity, as emotional commitment or intimidation? Critical listening may have to distinguish between (1) the message the communicator stresses, (2) the message the communicator mentions but does not stress, (3) the message that can be "read between the lines," and (4) the message the communicator might have delivered, but instead chose not to speak (Riggins, 1997; Starosta, 1973). Differences in cultural style most often are used to interpret the "messages between the lines," i.e., what the speaker "actually" was saying while not explicitly making those statements.

The untrained intercultural listener may wonder why some expected messages were not spoken, and why some anticipated stylistic cues were absent. In short, Culture One is deficient in its ability to deal with and to process messages from those of other cultural origins. Such deficiency, when it does not completely deter contact among culturally-different persons, still serves to distort those Culture Two messages that the listener receives.

Cultures also impose formal expectations on messages. The "secretary" of an Eastern European delegation does not equate to the "secretary" of a North American delegation, since the first usually holds significant power, while the latter typically records and schedules. The Swiss tend to go directly to the point, but East Asians and Central Americans may take more time to reach significant discussions. A North American speaker may stress his or her opinion on a matter, whereas the translation into Russian may leave out considerations of agency in order to concentrate on "the matter under consideration is ..." (Glenn, 1976). Chinese and Japanese are inclined not to directly say "no," for fear of damaging face, whereas Northern and Western Europeans often express their disagreement bluntly.

In media formats, American broadcasters simulate a friendly conversation with their audience, but British broadcasters maintain a formal, distinct separation from their audience. Western films vary in their degree of censorship, whereas Asian films ordinarily impose stricter codes of visual expression. Indian films last longer and contain many songs, but European films seldom contain songs and dance not integral to the plot. Culture Two listeners bring formal expectations to mediated and non-mediated cultural encounters by seeking gratifications that the messages are not designed to offer. They place aesthetic expectations and those of arrangement, stress, and presentation onto messages of other cultures that may dispose them not to want to hear and to process such messages. Processing messages that violate formal expectations takes significantly more energy than the processing of the more familiar formats of Culture One messages (Chen & Starosta, 1998).

In those cases where the Culture Two exchange is rhetorical, Hammerback and Jensen (1994) expand on Starosta and Coleman (1986):

1. The particular rhetorical form should be examined in relation to its rhetorical legacy in its ethnic culture;

2. The role of the specific rhetorical medium in that culture should be examined;

3. Any enumeration of the expectations should proceed from an understanding of cultural presuppositions as well as the rhetorical legacy of the form of communication;

4. The rhetor's or speaker's style of discourse should be viewed broadly to include the various structures, images, content, strategies, and appeals of a rhetorical piece;

5. In some cultures, the 'speaker's' image might not require much attention, both because speeches are not the primary means of rhetorical communication and because the speaker is in reality a group which wishes to remain unobserved (Hammerback & Jensen, p. 67).

This list may be extended to include rhetorical taboos. Speech forms play a designated role within cultures that spell out who may communicate (men but not women may be expected to speak to mass audiences on public occasions); that defines what topics are suited to public expression (Chinese would seldom publicly criticize officials or elders); that assigns propitious times and places to messages (some cultures dislike talking business over dinner); or that designates that criticism may follow ritualized drinking, but then must be forgotten (as for Japan [Critchfield, 1998]). Interethnic speech that violates such taboos could invite a harsh response, though it might gain an attentive audience because of its cultural novelty. On some rare occasion, a violation might be tolerated as a one-time exception, as when the Pope was permitted to speak and to hold mass in Cuba.

In sum, speakers wittingly or unwittingly represent a community or social grouping, they speak to those for whom some social identity is salient, their group and the group of the other may exhibit a smooth or strained history of interaction, the two groups may or may not share a common first language, historical power relations may have placed one party's group above the other, and recent events may or may not have transpired that acted as irritants between the two communities. Listening across ethnicities introduces ambiguities that seldom if ever appear in Culture One. The globalization of listening requires heightened tolerance for ambiguity, new awareness of expectations regarding proper settings and times for messages, and the willingness to treat some violations of communication norms as accidents, not as deliberate attempts at insulting another's cultural traditions.

Barriers to Global Listening

Culture One instills a reluctance to expose the self to the messages and meanings of the other. At a national level, governments may limit Culture Two contact, they may set a maximum percentage for foreign-originated films and television features, or they may decide which events get covered in the media or not. Some of their controls are enactments that impose censorship, whereas other controls consist of threats to employment or of limiting the allocation of newsprint to an offending newspaper. As one example, citizens of Saudi Arabia are generally expected to listen only to national radio stations, especially in the case of women, for fear the citizenry might hear messages that will go against the government's positions or against the teachings of Islam (Matawi, 1993). One barrier to global listening, then, is government efforts to limit Culture Two exposure.

National media tend to give the most attention to events that originate within, or directly concern, that nation. The debates surrounding the international flow of information illustrate the degree to which a few media centers originate the greatest share of the total media messages. Within those few countries not much is mentioned about other nations, especially those of the (developing) South, while the nations of the South use (metropole) Northern news sources to the neglect of their own national agendas. Selective reception of Culture Two messages is the second barrier to global listening.

Persons, organizations, and nations hold points-of-view. The media agendas in most nations closely parallel the policy agenda of those nations (Mahmoud, 1993) and news stories within the respective nations frame or slant the news in a direction that sanctions the actions of that nation (Starosta, 1971). Most or all events can be portrayed in a multitude of ways, in formulations that suggest a variety of motives and attitudes. The tendency to disparage the motivation of others represents the third barrier to global listening.

Linguistic communities within nations pose a barrier to listening. As an example, the Francophone press in Montréal and the local Anglophone press came to differ on the interpretation of a standoff in the summer of 1990 among Oka Mohawks and the police and military forces (Starosta & Hannon, 1997). The French-language press largely held that the confrontation was provoked by means of a Mohawk ambush, while the English-language media mainly portrayed the standoff as the use of excessive, provocative force by the Francophone police, in response to legitimate environmental and cultural concerns of the Mohawks. Those who listened to one or the other media Culture One explanations of events became alienated from members of the other linguistic community. The tendency of Culture One media to place parochial spins onto public events and controversies constitutes a fourth barrier to global listening.

New information grows exponentially. In the absence of someone to monitor and to select among possible messages, an auditor will not likely assign salience to a Culture Two concern. Listening requires attending to a message, considering where the message falls within a tapestry of related information, and assigning importance to the information that is being introduced. For the receiver to enter into a listening dialogue with the originator, motivation level must be high at the onset and must be maintained throughout the exchange. The difficulty of reaching such a level of motivation and intensity regarding Culture Two events is a fifth barrier to global listening.

Double-Emic Listening

> We must shift the organizing center of communicative activity so that it is neither in the individual nor in the linguistic system but is in the situation ... [this involves] a shift from a formal, instrumental toward an ethical stance: a concern with the being of others. For what is at stake is not simply the extent to which different voices may speak but the extent to which they will be responded to by those around them ... (Shotter, 1997, pp. 26–27).

Messages that are the product of global listening overlay elements from one culture onto those of the other. The new elements, for each party, are "figures" that are metaphorically grafted onto the "ground" of the familiar Culture One. Each party to the listening exchange will start from, and will favor, its Culture One ground, and will consider the new information to be a point-of-view, i.e., something less than fact.

For some amount of time the perspectives of the more powerful party will be used to interpret events. Those with the greater power need not define themselves as a perspective or as a point-of-view, since they see themselves as the truth and as the center. Authentic global listening requires those of the majority or with the best-established media to problematize their own culture, to critically examine it, and to see it as only one possible Culture One.

The act of problematizing one's own Culture One perspective creates the possibility of becoming open to alternative understandings. The outcome of global listening is neither the culture of the one party nor of the other; it lies between the two possibilities.

The view of the world as seen by one who lives it may be described as an "emic" view. The view of one's world as seen from a viewpoint outside the culture may be thought of as an "etic" perspective. Global listening moves the interacting parties away from their moorings to develop a space-between. The parties soon move from their respective emic views to formulate an etic view of the other. Global listening finally carries the listeners to a double-emic state in which they construct a new reality from pieces of their own Culture One and pieces of the Culture One of the other. Double-emic listening takes place among interactants who are aware that their own Culture One is a theory; that another's views could provide valid alternative views; and that information from the other is needed to grow and to survive as a global citizen.

CONCLUSION

Listening among those of differing cultures moves from an intracultural stage (Culture One) to an interpersonal stage (Culture Two). It then enters an intercultural stage of relationship building and restructuring, and a rhetorical stage of internalizing any changes. Finally, it enters a new intracultural stage at a more global level. The rhetorical stage is one in which variant values are negotiated and adjusted among parties so as to render them mutually palatable. The restructured view of the world that follows global listening can be termed Culture Three.

Research is required on listening across diversity in global society to see how the process of presentation, interaction, adjustment, and internalization can be made to function more smoothly. Movement from Culture One through Culture Two to Culture Three should be carefully researched and documented to learn how global listening can promote diversity, mutual learning, and peaceful interaction among nations and peoples.

NOTES

1. Culture One is a construct the first author has used as a teaching device for 241–615 Intercultural and Interracial Communication, Howard University, to link the various approaches taken by researchers under the rubric "intercultural communication." Five "Cultures" are offered within this schema.

2. The analysis is adapted from Starosta, 1997.

3. Orbe proposes that minority cultures or those with little power must devise communication strategies in order to deal with the mainstream culture. By contrast, the present analysis proposes that even the majority culture be viewed as a co-culture, lest this paper should reify and essentialize the very power disparity that Orbe so compellingly interrogates.

4. This we term Third Culture Building. It may arise with or without the urging of facilitators and detractors.

REFERENCES

Bakhtin, M.M. (1981). *The dialogic imagination.* (C. Emerson & M. Holquist, trans.). Austin: University of Texas Press.

Barthes, R. (1971). *Image, music, text* (S. Heath, trans.). New York: Hill and Wang.

Bormann, E. (1972). Fantasy and rhetorical vision: The rhetorical criticism of social reality. *Quarterly Journal of Speech, 58*, 396–407.

——. (1980). *Communication theory.* NY: Holt, Rinehart & Winston.

——. (1982). Fantasy and rhetorical vision: Ten years later. *Quarterly Journal of Speech, 68*, 288–305.

Chen, G.M., & Starosta, W.J. (1998). *Foundations of intercultural communication.* Boston: Allyn & Bacon.

Critchfield, A.J. (1998). *Report on Japanese communication, 12*, October 1998.

Farrell, T.B. (1993). On the disappearance of the rhetorical aura. *Western Journal of Communication, 57*, 147–158.

Fletcher, C. (1992). The semiotics of survival: Street cops read the street. *The Howard Journal of Communications, 4*, 133–142.

Glenn, E.S. (1976). Meaning and behavior: Communication and culture. In L.A. Samovar & R.E. Porter (eds.) *Intercultural communication: A reader*. Belmont, CA: Wadsworth.

Hammerback, J.C. & Jensen, R.J. (1994). Ethnic heritage as rhetorical legacy: The Delano Plan. *The Quarterly Journal of Speech, 80*, pp. 53–70.

Huspek, M. (1997). Communication and the voice of the Other. In M. Huspek and G.P. Radford (eds.) *Transgressing discourses: Communication and the voice of the other.* Albany, NY: SUNY Press.

Mahmoud, A.A. (1993). *A time-series analysis of the portrayal of Palestinians by selected U.S. and West European press from 1948 to 1988.* Unpublished diss., Howard University.

Matawi, A. (1993, April). *Patterns of Radio Listenership to Foreign Radio Broadcasts in Western Saudi Arabia during the Gulf War.* Unpublished M.A. thesis, Howard University.

Orbe, M. (1998). Constructing co-cultural theory. Thousand Oaks, CA: Sage.

Riggins, S.H. (1997). (ed.) *The language and politics of exclusion: Others in discourse*. Thousand Oaks, CA: Sage.

Shotter, J. (1997). Textual violence in academe: On writing with respect for one's others. In M. Huspek and G.P. Radford (eds.)

Transgressing discourses: Communication and the voice of the other. Albany, NY: SUNY Press.

Starosta, W.J. (1971). United Nations: Agency for semantic consubstantiality. *Southern States Speech Journal, 36*, 243–254.

——. (1973). Non-communication as game in international relations. In M.H. Prosser (ed.) *Intercommunication among nations and peoples*. New York: Harper & Row.

——. (1997). *Six propositions on news narratives: Reporting on interethnic conflict.* Paper presented at World Communication Association, San José, Costa Rica.

Starosta, W.J. and Coleman, L. (1986). A Case Study of Rhetorical Interethnic Analysis: Reverend Jackson's 'Hymietown' Apology," in Young Yun Kim (ed.) *Interethnic Communication: Current Research.* Newbury Park, CA: Sage.

Starosta, W.J. and Hannon, S.W. (1997). The multilexicality of contemporary history: Recounted and enacted narratives of the Mohawk Incident in Oka, Québec. *The International and Intercultural Communication Annual, 20*, 141-165.

CHAPTER SIXTEEN

Implications of Communication and Globalization for Future Research: End Remarks

Barbara S. Monfils
University of Wisconsin, Whitewater

> Take an individual out of his own culture and set him down suddenly in an environment sharply different from his own, with a set of different cues to react to—different conceptions of time, space, work, love, religion, sex, and everything else—then cut him off from any hope of retreat to a more familiar social landscape, and the dislocation he suffers is doubly severe. Moreover, if this new culture is itself in turmoil, and if—worse yet—its values are incessantly changing, the sense of disorientation will be still further intensified. Given few clues as to what kind of behavior is rational under the radically new circumstances, the victim may well become a hazard to himself and others.
>
> Now imagine not merely an individual but an entire society, an entire generation—including its weakest, least intelligent, and most irrational members—suddenly transported into this new world. The result is mass disorientation, future shock on a grand scale.
>
> This is the prospect that man now faces. Change is avalanching upon our heads and most people are grotesquely unprepared to cope with it.
>
> — Toffler

INTRODUCTION

When Toffler wrote these words in 1970, he described a world in which the pace of change was accelerating so quickly that human beings would be incapable of adjusting to the changes. Driven by technology, people and cultures would suffer "future shock," as they were no longer able to keep up with the multiplicity of changes occurring throughout society.

Not all of Toffler's predictions have come to pass. However, he was perceptive in predicting the far-reaching effects of technology on nations and cultures. In the contemporary world, this has been manifested through globalization. As the new millennium takes shape, communication will be a driving force in defining the global community in the 21st century. Each of the authors of articles in this volume has noted ways in which s/he believes communication will shape the future. These forces will, in turn, converge and diverge to influence and shape the nature of communication and cultures. Although not the only issues, four major issues will be especially important for communication researchers interested in globalization: definitions of community; politico-economic forces as driving forces; language; and citizenship. Each of these will be refined and molded by technologies of communication.

Before each of the four major issues above can be analyzed, it is necessary to discuss a starting point for many of the essays in this volume: the importance of the technologies of communication in the development of global perspectives in communication. The growth in technology and globalization are important reasons for studying intercultural communication, according to Chen & Starosta (1998), Lustig and Koester (1999), Klopf (1998), and Martin and Nakayama (1997). Whether used as an assumption (e.g., Shuter in this volum) or a focus (e.g., Cooks; and Ma in this volume), for analysis, the authors of the essays in this volume accept that whatever impact globalization will have on peoples and cultures, one of the principal shapers is and will continue to be communication technology. Media cultures as formulated by Chesebro (1989) have resulted in both real and "virtual" dimensions of change and development. For some authors in this text, the issue becomes how communication technologies are redefining existing social roles and institutions; for others, the issue is how communication technologies will create new communities and citizens. Still others warn about the consequences, including division, marginalization, and conflict, in the new millennium. Thus, a discussion of each of the four issues must be predicated on the assumption that information technologies will have a vital role in how each of these dimensions evolves and shapes society, cultures, and peoples.

It is also important to define the term, "globalization." Although different researchers use the term to describe a myriad of characteristics associated with the term, Lynch (1998), quoting Friedman, identifies core elements of globalization: "This consensus [on the definitional core of globalization] is perhaps best expressed by a prominent U.S. journalist [Friedman], who has defined globalization as 'that loose combination of free-trade agreements, the Internet and the integration of financial markets that is erasing borders and uniting the world into a single, lucrative, but brutally competitive, marketplace'" (p. 149). Thus, the four major issues discussed in this essay: community, politico-economic factors, language, and citizenship, are all components that must be taken into consideration when discussing globalization.

THE GLOBAL COMMUNITY

Among communication researchers, there seems to be consensus that the "global village" predicted by McLuhan is coming to fruition. Communication technologies have linked persons and cultures together in ways unforeseen in past centuries. It is no wonder, then, that new and different definitions of community are evolving.

One set of terms used to describe communities in the information age focuses on the influence of the communications media themselves. "Wired cities" (Dutton, Blumler & Kraemer, 1987), "virtual communities" (McChesney, 1996), and in this volume, "cyber-communities" (Cooks), "global community" (Holt), and "global society" (Chang) are among the terms that have been used to refer to the linking of persons into virtual communities by means of technology. In these terms, previously-used definitions of community, including physical proximity, shared norms, kinship patterns, linguistic similarity, and sense of security are no longer adequate definitions for what creates a sense of community. Mass media, ranging from short-wave radio to the Internet, allow people to surpass geographical boundaries in their communications with others. Neighbors co-exist with neighbors without knowing the names of the person who lives next door. For many people in the global community, recreational time is as likely to be shared in game playing in virtual reality as it is in a playing field, park, or village square (Real, 1996). Mobility has become more and more a fact of life as the promise of long-term employment is being replaced by policies of "downsizing," "right-sizing," and "outsourcing." In communities that are becoming increasingly diverse, expectations about the use of a monolingual tongue are being supplanted by contemporary versions of "trade languages." At the same time, shifting migrations patterns in the U.S. are reshaping notions about the appropriateness of maintaining one's own language while living in an environment in which a different language is commonly spoken. At times the debates are conducted in governmental circles, such as in the "English-only" debates that have taken place in the legislatures of more than forty states in the United States and in Congress within the last ten years. Other times, educational institutions bear the brunt of the attention, as evidenced by controversy surrounding the proposal to teach Ebonics in Oakland, California, in 1996. Sometimes the debate takes on ideological and political implications, as in the 1998 plebiscite in Puerto Rico, in which voters selected "none of the above" in greater numbers than statehood, commonwealth (based on how that term was defined on the ballot), or independence. The 1998 plebiscite was the fifth election held on the island since it became a commonwealth. In this campaign, opponents of Puerto Rico's becoming a state contended that a vote for statehood would mean the imposition of English as the only "official" language on the island, which was governed by Spain for more than 400 years before being ceded to the U.S. in 1898.

Finally, a sense of security can no longer be used as a means of defining community. If the atomic bomb shattered humanity's sense of security in the mid-twentieth century, the growth of terrorist tactics in the latter decades of the century have further contributed to a sense that no matter what people may try to do, they cannot eliminate the element of risk. From bombs, car bombings, and airlines that disappear from radar screens in an instant, human beings are increasingly being made aware that no matter what precautions they themselves may take for protection, they are always vulnerable. Moreover, places that in times past were viewed as secure have lost that designations, as evidenced by terrorist attacks on tourist attractions (United States), subways (Japan), government buildings (Colombia; United States), embassies (Kenya and Tanzania), and the residences of ambassadors (Peru). When these attacks occur, they generally occur with deadly force and produce casualties. Since no one is safe, it is no wonder that strangers are viewed with suspicion, even when their motives seem harmless enough.

Yet, these points of divergence coalesce at a deeper level to converge into a basic human need—the need to be around other people. Human beings are, by nature, social animals; as human communicators, we learn about and continue to sustain ourselves through our interactions with others. In short, we need to communicate with others to become fully human. Thus, as we lose our "old" ways of defining community, they must be replaced with new methods of defining the "we-ness" that is the hallmark of community.

Information technology provides one means of redefining community. By maintaining contact with traditional "roots" through the use of media, a sense of community can be established and/or maintained. In this volume, Cooks' example on the use of the Internet by Panamanian residents provides one example by which a link to one's ethnic, cultural, or national background may be maintained. Ma's analysis of the Internet as a virtual "town square" is another. Chen &

Starosta's description of an inclusive "epistemic" community is a third. Although no one can argue with certainty the specific components that will define "community" in global society, there is consensus that such a community is in the process of emerging.

POLITICO/ECONOMICS FORCES

In some ways, the driving forces of politics and economics may seem to be separate entities. However, the merging of political and economic forces that have characterized the latter part of the twentieth century also provide a rationale for their use in combination. Tenranian (1998) has labeled the emerging global political economy as "pancapitalism." Among the characteristics that Tenranian cites are the change from industy-based to information-based economies, the move to flexible accumulation of goods and services, the transition from national to global markets, and the growth of "gated" communities. Citing statistics from the 1996 UN Human Development Report, Tenranian documents the widening "gap" between the "haves" and the "have nots" in the emerging global political economy:

> The poorest 20 percent of the world's people saw their share of global income decline from 2.3 percent to 1.4 percent in the past 30 years. Meanwhile, the share of the richest 20 percent rose from 70 percent to 85 percent. That doubled the ratio of the shares of the richest and the poorest—from 30:1 to 61:1 ... The assets of the world's 356 billionaires exceed the combined annual income of countries with 45 percent of the world's people (pp. 83–84).

Tenranian envisions one of three possible scenarios as nations and peoples cope with the new economic realities: continuity, collapse, or transformation. Tenranian further agues that a new set of international ethics will be needed, and that intercultural learning is a necessary part of the development of new ways of thinking, learning, and communicating. Several authors, such as Chen, Chung, Heisey, Larson, Wildermuth, and Zhong, in this volume echo Tenranian's points.

One of the major economic factors contributing to globalization in the last several decades has been the international expansion of organizations. Intercultural communication scholars have noted that both the organization and the national culture have been influenced by multinational organizations. As the number of multinational corporations has grown, so, too, have charges of economic and cultural imperialism. Berger (1997) uses terms such as "Davos culture," "faculty club international," "the McWorld culture," and "Evangelical Protestantism" in detailing four interpretations of the cultural revolutions precipitated by global economic forces. Although not referring specifically to Berger's terms, several authors in this volume have addressed the same issues. In his essay, for instance, Chuang provides several examples to demonstrate the consequences of "Americanization" or "Russification," and Shuter offers examples both in terms of numbers and names of multinational communications industries and their potential for impacting globalization. Indeed, this area seems to be one that has generated much research interest.

The impact of economic globalization has affected political and social trends as well. In some cases, economic forces have had the impact of realigning boundaries that have transcended the traditional boundaries of nation-states (Frederick, 1993). The growth of transnational alliances, such the Organization of Petroleum Exporting Countries (OPEC), in the 1960s and 1970s, demonstrated that when geographically-separate countries with a common product acted as a unit, they could significantly impact the economies of other nations. The strong economies of the individual nations comprising the "Five Asian Dragons" of Hong Kong, Singapore, Taiwan, Japan, and South Korea have had a similar impact on the economies throughout Asia in the late 1980s (Chen & Chung, 1994). The passage of the North American Free Trade Agreement (NAFTA) in 1993 has not only resulted in the reduction of trade barriers among the United States, Canada, and Mexico, but has been extended into South American nations such as Chile as well. Mexican border towns have also been significantly impacted by the *maquiladoras* that have been established since NAFTA went into effect (McDaniel & Samovar, 1997). Finally, fifty years ago, only the most foolhardy of seers would have predicted that the European Community would not only form a union for the reductions of customs and trade barriers, but would also move toward the implementation of a single currency. Yet on the 1st of January 1999, banks throughout Europe began conducting financial transactions using the Euro currency. Chang's assertion in this volume that the creation of a single market system among China, Taiwan, and Hong Kong could result in a "Greater China" is but one example of the conflicting forces of divergence and convergence in politico-economic globalization. In the global marketplace, communication will be the "cement" that binds politico-economic forces. The nature of that "cement" will provide much data for future investigation.

However, the media have not been neutral in driving these politico-economic forces. The Western, market-driven model of economic development promoted by the media is influencing peoples and cultures throughout the world in subtle but significant ways. Examples of nations who have changed their economic strategies to a market-driven model in the latter decades of the 20th century abound, from Asian countries, including China and Russia, to many Eastern European nations to South American nations such as Chile, Brazil, and Peru. In each case, economic changes had (and are continuing to have) consequences for the cultures and societies of those nations. Thus, one cannot examine politico-economic forces and globalization without necessarily analyzing the impact of the media, particularly its implicit promotion of the market-driven economic model of development, Western in nature, on forces and factors of globalization. In citing Oliveira's 1991 study of Brazil in his essay, Shuter further illustrates these concerns. Though the focus of research is different, the underlying issues Shuter raises are not unlike those identified by Protzel (1996), in his analysis of how the mass media contributed to the fostering of a sense of national identity in Peru in the 1980s and 1990s.

In the essay for this volume, Zhong argues that the end of the "Cold War" may well have been the catalyst that allowed nations to focus more on their own economic development than on matters of defense. Given the nature and scope of political changes that have occurred in many nations in Asia, Africa, and Latin America in the 1990s, there is little doubt that the 1990s have been a decade of profound political and economic change (Gonzalez-Menet, 1988). These political and economic changes, spurred on in large part by information technologies, have provided opportunities for economic growth and development, while at the same time, have made life more difficult for those "left behind." The political, social, and cultural changes that will inevitably occur in the emergent global economic system will provide a unique set of areas for study by communication researchers in the years to come.

LANGUAGE

Just as political and economic forces have changed the composition and construction of the world, language has provided a catalyst for points of convergence and divergence as well. Once again, arguments are being reshaped to respond to forces of globalization.

When studying linguistic phenomena in the future, linguists may well identify the ascendance of English as the major linguistic hallmark of the 20th century. The expansion of the economy of the United States, both internally and externally, has resulted in the proliferation of English as the language of trade and business throughout much of the world (Asuncion-Lande, 1998; Stevenson, 1994; Berger, 1997). Moreover, the growth of computer technology, the foundation of which has been largely developed in English, has given rise to the use of English as a medium of technical communication. The assertion of the United States as the "giant" in military/diplomatic communication has resulted in the replacement of French by English as the language of diplomacy. Although Chinese is the first language of one-quarter of the world's population, there are now more speakers of English as a second language in China than there are native speakers of English in the world. Thus, arguments of "linguistic imperialism" are being raised by those who feel that the widespread use of English is not only a basis for economic exploitation, but for linguistic, social, and cultural exploitation as well (Stevenson, 1994).

Language is not a neutral entity; those who speak the language are influenced by the nuances inherent within that language. Such influences can range from a choice of words to embedded cultural assumptions about the nature of humanity, existence, and human responsibilities (Berger, 1997). If Veliz, as quoted in Berger, is correct in his assertion that we live in "a world made in English," then those who study the interrelationships of language and communication have a vast area of research for some time to come.

Questions of language use are made even more complex when issues of marginalization are considered. As globalization continues, what will happen to those who do not speak one of the "major" or "trade" languages? Will their concerns be voiced, and who will be the spokespersons for their concerns if they are not able to make themselves understood by others? These questions are especially significant in Africa, in which as many as 2,000 indigenous languages are currently spoken (Ndiokwere, 1998). As Africans struggle with the co-existence of tribal, national, and "official" languages, how much further behind will those be who do not have the access or the means by which to learn the global "trade" languages of the 21st century? For decades, development in many African nations has been hampered by the "brain drain" as many of its elites have sought better opportunities outside of their own countries. Though not unique to Africa, the "brain drain" phenomenon widens the "gap" between those who have access to information technology and development and those who will be left behind. Indeed, marginalization due to language may well become one of the most significant points of contention as globalization continues to evolve.

A related set of issues focuses on the linguistic experiences of those who choose to emigrate to another culture. Historically, immigrant populations established themselves in their new environment and went to work as unskilled laborers. They sent their children to school so that their children would be fluent in the language when they entered the workforce. Thus, by the time the children are ready to enter the workforce, they are able to communicate in the language required in the business environment. In today's service- and information-based economy, however, the demand for workers who do not speak the language is significantly reduced. Thus, paradigms that previously accounted for the integration of immigrants into a new cultural and linguistic milieu are no longer adequate. The United States, for instance, is currently undergoing a transformation due partially to the large influx of non-English-speaking persons who have immigrated in recent years. Long-term effects are only beginning to be documented, much less studied, and will form the basis for much scholarly research in language and communication in the workplace for years to come.

A final set of issues focuses on language and communication at the level of policy. These questions will range from the language of instruction in elementary and secondary schools to issues of privacy and free speech. As evidenced by the controversy surrounding California's Proposition 227 in 1998, attitudes toward the need to speak English to fully participate in the United States are deeply rooted. Does this mean, however, that English should be the exclusive language of communication? In fact, given the demographic changes expected to occur in the United States in the next several years, proponents and opponents to "English only" can be expected to become more vocal in expressing their views. Questions of language policy will thus be raised that will hit at the "heart" of the study of communication. Will the study of oral communication become less important as more information is communicated digitally? Will the need for fluency in oral communication become increasingly important as the sheer volume of information continues to grow? As volume of available information expands, who are the gatekeepers? And who will be ascribed as taking responsibility when information once considered offensive is disseminated and posted on "public" information networks, such as the Internet? In global society, monolingualism will not be sufficient; which languages provide avenues for full participation, particularly in light of the rise of English as a worldwide "trade" language, have yet to be decided.

CITIZENSHIP

A final set of issues focuses on questions of citizenship in the global community. Kegley (1997) argues that the tendency to study international affairs from an interdependent, pan-national perspective, rather than a state-centric perspective, has been a recent phenomenon in world history. Thus, new sets of perspectives emerge when considering citizenship outside of one's identity as stamped on a passport. While national boundaries will not disappear, many of the authors in this volume establish that one's own nationality is but one means of defining the global citizen. While no one knows what other components will comprise the global citizen, several predictions have been made. Each of the authors' predictions contains points of convergence and divergence.

Chen and Starosta confront citizenship in the global community directly in their definition of the global citizen. Arguing that such citizenship must be cultivated, they posit a sense of inclusiveness in their definition. Their sense of community completes a circle as it brings us back to the first issue discussed in this essay—what the nature and scope of the global community will be like. Nonetheless, they are optimistic that as citizens, people will take their place in the transformed communities of tomorrow. Hence, another point of convergence is established.

Cooks, too, develops a definition of community in her essay, illustrating predictions through examples. Cooks' description, however, is more two-sided. She not only discusses the components of citizenship, but also brings up one of the dilemmas of citizenship in her essay: the marginalization of out-group members. Thus, issues of language and politico-economic factors are implied in her view of citizenship in the global community.

Chang discusses citizenship in the context of a "greater China." Implicit in this is the assumption that geographical boundaries can become transparent within certain contexts, such as historical events or ethnic identification. In such a community, more long-standing circumstances provide a meta-umbrella through which geographical boundaries, established by recent political events, are dissolved into the background.

Finally, citizenship may emerge from a synthesis of other identities, as argued by Chuang. Through the promotion of a sense of internationalism, the citizens of the global community may find themselves as gaining a new sense of citizenship that incorporates elements of several other types of citizenship. This may be viewed as analogous to a third culture perspective, in which the individual selects those elements from several definitions and molds a new definition. While having some similarities with other types of citizenship, this type is nonetheless unique. In contemporary society, according to Chuang, internationalism will be a necessary component. When internationalism is raised as an issue, politico-economic factors become prominent sub-texts in defining citizenship.

Thus, while various authors provide highly divergent predictions as to the nature and scope of citizenship in the global community, their predictions converge in one inescapable fact: humans must learn to live as citizen-communicators in the global community. The solution to living in peace in spite of the inevitable conflicts that will be developed in pan-capitalist global society, argues Tenranian (1998), include: "… a form of communication that is dialogical in character and transnational and inter-civilizational in its epistemological reach (p. 85)." As such, there must be a set of ethnical principles by which human behavior in the global community may be assessed. As Shuter points out in his essay, however, the establishment of a single ethical system in the global environment will not be easy to establish. Indeed, this may pose the single greatest challenge for the future.

CONCLUSION

As the 20th century comes to an end, Toffler's dire predictions of a world in disarray due to "future shock" have not occurred on a widespread scale. Nonetheless, the nature of community, politico-economic factors, language, and citizenship have emerged as major issues as nations and people try to adjust to a world that is becoming more and more global in nature. These factors impact communication at all levels. As human beings, we are all participant-observers in identifying, understanding, and interpreting how communication will change and be changed by processes of globalization. As communication researchers, we have the additional responsibility of studying these processes as they occur, and of documenting their effects on peoples and cultures. This volume is one voice in dialogue that is certain to continue for decades to come.

REFERENCES

Asuncion-Lande, N (1998). English as the dominant language for intercultural communication: prospects for the next century. In K. S. Sitaram & M. Prosser (Eds.), *Civic discourse: Multiculturalism, cultural diversity, and global communication* (pp. 67–82). Stamford, CT: Ablex, pp. 67–82.

Berger, P.L. (1997). Four faces of global culture. *National Interest, 49*, 23–30.

Chen, G. M., & Chung, J. (1994). The 'five Asian dragons': management behaviors and organizational communication. *Communication quarterly,42*, 93–105.

Chen, G. M., & Starosta, W. J. (1998). *Foundations of intercultural communication*. Boston: Allyn & Bacon.

Chesebro, J.W. (1989). Text, narration, and media. *Text and performance quarterly*, 9, 1–23.

Dutton, W.H., Blumler, J.G., & Kraemer, K.L. Eds. (1987). *Wired cities: shaping the future of communications.* Boston: G.K. Hall.

Frederick, H.H. (1993). *Global communication & international relations.* Belmont, CA: Wadsworth.

Gonzalez-Manet, G. (1988) (L. Alexandre, Tr.). *The hidden war of information*. Norwood, CT: Ablex.

Kegley, C. (1997). Placing global ecopolitics in peace studies. *Peace review, 9*, 425–430.

Klopf, D. W. (1998). Intercultural encounters: The fundamentals of intercultural communication. Englewood, CO: Morton Publishing Co.

Lustig, M.W., & Koester, J. (1999). Intercultural competence: interpersonal communication across cultures, 3rd ed. New York: Longman.

Lynch, C. (1998). Social movements and the problem of globalization. *Alternatives: social transformation and humane governance, 23*, 149–174.

Martin, J. N., & Nakayama, T.K. (1997). *Intercultural communication in contexts*. Mountain View, CA: Mayfield Publishing Co.

McChesney, R.W. (1996). The Internet and U.S. communication policy-making in historical and critical perspective. *Journal of Communication, 46*, 98–124.

McDaniel, E.R. & Samovar, L.A. (1997). Cultural influences on communication in multinational organizations: The maquiladora. In L. Samovar & R. Porter (Eds.), *Intercultural communication: a reader* (pp. 289–296). Belmont: Wadsworth, pp. 289–296.

Ndiokwere, N.I. (1998). *Search for greener pastures; Igbo and African experience*. Kearney, NE: Morris Publishing.

Protzel, J. (1996). Terror, media and disbelief in Peru. In D. Paletz (Ed.), *Political communication in action* (pp. 73–96). Cresskill, NJ: Hampton.

Real, M. (1996). *Exploring media culture: A guide.* Thousand Oaks, CA: Sage.

Stevenson, R.L. (1994). *Global communication in the twenty-first century*. New York: Longman.

Tenranian, M. (1998). Dialogues of civilizations for peace. *Peace review, 10 (1)*, 83–88.

Toffler, A. (1970). *Future Shock.* New York: Bantam Books.

INDEX

T

U

V

W

X

Z